Modern Birkhäuser Classics

Many of the original research and survey monographs in pure and applied mathematics published by Birkhäuser in recent decades have been groundbreaking and have come to be regarded as foundational to the subject. Through the MBC Series, a select number of these modern classics, entirely uncorrected, are being re-released in paperback (and as eBooks) to ensure that these treasures remain accessible to new generations of students, scholars, and researchers.

Knot Theory
&
Its Applications

Kunio Murasugi

Reprint of the 1996 Edition
Birkhäuser
Boston • Basel • Berlin

Kunio Murasugi
Department of Mathematics
University of Toronto
Toronto, Ontario M5S 2E4
Canada

Originally published as a monograph.

Cover design by Alex Gerasev.

Mathematics Subject Classification (2000): 57M25, 57-01

Library of Congress Control Number: 2007933901

ISBN-13: 978-0-8176-4718-6   e-ISBN-13: 978-0-8176-4719-3

Printed on acid-free paper.

©2008 Birkhäuser Boston   *Birkhäuser*   🅑®

9 8 7 6 5 4 3 2 1

www.birkhauser.com   (IBT)

Kunio Murasugi

Translated by Bohdan Kurpita

KNOT THEORY
and
ITS APPLICATIONS

Birkhäuser
Boston • Basel • Berlin

Kunio Murasugi
Department of Mathematics
University of Toronto
Toronto, Ontario M5S 1A1
Canada

Bohdan Kurpita
The Daiwa Anglo-Japanese
Foundation and Waseda University
Shinjuku-ku, Tokyo
Japan

Library of Congress Cataloging-in-Publication Data

Murasugi, Kunio, 1929-
   [Musubime riron to sono ōyō. English]
   Knot theory and its applications / Kunio Murasugi ; translated by
Bohdan Kurpita.
     p.   cm.
   Includes bibliographical references (p. -) and index.
   ISBN 0-8176-3817-2 (alk. paper). -- ISBN 3-7643-3817-2 (alk.
paper)
   1. Knot theory.   I. Title.
QA612.2.M8613   1996   96-16329
514'.224--dc20   CIP

Printed on acid-free paper
Published originally in 1993 in Japanese   *Birkhäuser*
© 1996 Birkhäuser Boston

ISBN 0-8176-3817-2
ISBN 3-7643-3817-2
Typeset in T$_E$X by Bohdan Kurpita
Printed and bound by Maple-Vail, York, PA
Printed in the U.S.A.

9 8 7 6 5 4 3 2

Contents

Preface

Knots and braids have been extremely beneficial through the ages to our actual existence and progress. For example, in the primordial ages of our existence, in order to construct an axe a piece of stone was bound/knotted to a sturdy piece of wood. To make a net, vines or creepers, animal hair, *et cetera* were bound/braided together. Also it is known that the ancient Inca civilization developed a system of characters that were formed from knotted pieces of string.

Although people have been making use of knots since the dawn of our existence, the actual *mathematical* study of knots is relatively young, closer to 100 years than 1000 years. In contrast, Euclidean geometry and number theory, which have been studied over a *considerable* number of years, germinated because of the cultural "pull" and the strong effect that *calculations* and *computations* generated. It is still quite common to see buildings with ornate knot or braid lattice-work. However, as a starting point for a study of the mathematics of a knot, we need to excoriate this aesthetic layer and concentrate on the *shape* of the knot. Knot theory, in essence, is the study of the geometrical apects of these shapes. Not only has knot theory developed and grown over the years in its own right, but also the actual mathematics of knot theory has been shown to have applications in various branches of the sciences, for example, physics, molecular biology, chemistry, *et cetera*.

In this book, we aim to guide the reader over the multifarious aspects that make up this theory of knots. We shall, in a straightforward manner, explain the various concepts that form this theory of knots. Throughout this book, we shall concentrate on lucid exposition, and the exercises that can be found liberally sprinkled within act as a conduit between the theory and the understanding of this theory by the reader. Therefore, this book is not just another book for those who work or intend to work within the confines of knot theory, but is also for those engaged in other areas in which knot theory may be applied even if they do not have a considerable background in mathematics. The general reader is also welcome, hopefully adding to the diversity of knot theory. We shall cover what exactly knot theory is; what are its motivations; its known results and applications; and what has been discovered but is not yet completely understood.

Knot theory is a branch of the geometry of 3 dimensions. Since three dimensions is the limit of what is usually perceived intuitively, we can call on this to help us explain concepts. To this end, in this book we make extensive use of the numerous diagrams. Moreover, often the intuitive approach is carried through into the actual text. However, the proofs are still proven to the usual standard of mathematical rigour. In certain cases, for the convenience of the reader, we have appended at the end of this book several short, more detailed notes and commentaries.

Since we have tried as much as possible to avoid formal terminology, i.e., we do not use concepts that are common in topology, such as knot group and homology group, it has been necessary to leave out several theorems and proofs. For the reader who is interested in a more formal approach, a good guide is the book by Crowell and Fox [CF*]. For those interested in obtaining an even deeper understanding of knot theory than that which may be garnered by reading this book only, we recommend the research-level book edited by Kawauchi [K*]. Since the purpose of the bibliography at the end of this book is to cite the theorems that appear in the text, as a general bibliography for knot theory it is inadequate. However, in Kawauchi [K*] and Burde and Zieschang [BZ*] there can be found exhaustive bibliographies, so the inquisitive reader who requires further references should consider consulting these two bibliographies.

As a supplement, we include the knot diagrams of prime knots with up to eight crossing points (35 in total), and a second table lists their Alexander and Jones polynomials. Hopefully, this will prove of practical use to the reader.

Finally, during the gestation period of this book I received the valuable opinions of M. Sakuma, M. Saito and S. Yamada. Also, several people kindly explained ideas to me that are outside my field of speciality. The students of M. Sakuma and S. Suzuki provided many additional, helpful comments about the original Japanese edition. To all these individuals I express my deep gratitude. Furthermore, for the Japanese language edition of this book I received much help from the editorial staff of the publishers, especially from T. Kamei. For the English edition, the staff at Birkhäuser in Boston, especially E. Beschler, have been extremely helpful. To all these people I express my warmest thanks.

Postscriptum, even in the few years since the Japanese version was published in 1993 there have been interesting developments in knot theory. In this English translation, we have incorporated some of these recent developments.

Introduction

With a reasonably long, say 30cm in length, piece of string or cord, loosely bind a box as shown in Figure 0.1(a). You should now be holding in your hands a *simple* type of knot. Now take the two ends and glue them together so that it is not immediately noticeable that the string/cord has been joined. This exercise should be performed in such a way that the string does not come into contact with the box. The box is more a prop than a necessity. When the exercise is completed, what you should see before you is a single knotted loop, approximately 30cm long, Figure 0.1(b). In mathematics this loop is called a *knot*.

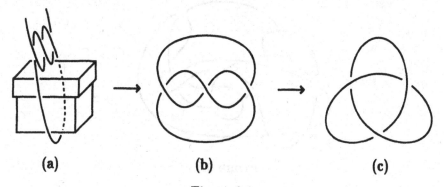

(a)   **(b)**   **(c)**

Figure 0.1

To be a bit more precise, this (slender) string should, ideally, be thought of as a single curve, then a knot is a simple closed curve; in space. If the reader is left-handed then the above knot will differ slightly in appearance and will take the form shown in Figure 0.2.

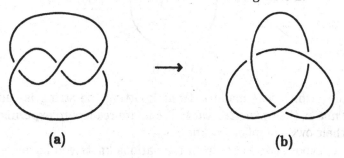

(a)   **(b)**

Figure 0.2

At first sight, the above two knots seem similar; however, with a more careful perusal of these two figures, it is possible to see that they differ in several places. These two types of knots are each called a *trefoil knot*, or sometimes due to their resemblance to a clover-leaf, a *clover-leaf knot*. Since the form of the trefoil knot differs in Figures 0.1(c) and 0.2(b), we on occasions in order to distinguish them, refer to the knot in Figure 0.1(c) as the *right-hand* trefoil knot and that in Figure 0.2(a) as the *left-hand* trefoil knot.

To describe a knot it is not sufficient just to say a knot is the object obtained when we bind a box or something similar with a piece of string. We can make various types of knots that are independent of how we tie the string. Let us create an extremely complicated knot by using a *very* long piece and knotting it in the most muddled fashion we can imagine; an example, due to Ochiai is given in Figure 0.3.

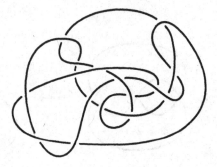

Figure 0.3

In contrast, *just* glue together the ends of a 10cm piece of string, as illustrated in Figure 0.4.

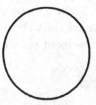

Figure 0.4

This knot, which was made *without* knotting the string is called the *trivial knot* or the *unknotted knot*. We invite readers to experiment and create their own complicated knots.

Now, choose two knots from the various knots you have created in this random manner and call them A and B. The natural question to

ask is, Can we change knot A into knot B? An approach we may try
is to determine how it is possible to change the knot in Figure 0.1(b)
to that in Figure 0.1(c). One condition we impose is that we cannot
cut the knot, we can only manipulate it by hand. If we can change A
into B by slowly changing the form of A, then these two knots are said
to be *equivalent* or *equal*. (This notion of equivalence is probably the
most intuitive; we give a more mathematical definition in Chapter 1,
Section 2.)

So, can we resolve the four knots in Figures 0.1(b), 0.2(a), 0.3,
and 0.4 into classes of equivalent ones? Before we attempt this, let us
further restrict our notion of a knot. Since what we are interested in
is the *actual form* of the knot, we do not need to worry about how
long or how thick the knot is. Let us mull a bit over the way we wish
to consider a knot. It is perfectly possible that a knot made from a
30cm piece of string is equivalent to a knot made from a 10cm piece of
string. Recall, we want to say that two knots A and B are equivalent
if we can manipulate, using our hands, the knot A into the knot B (to
visualize this it is best to suppose the strings have a certain amount
of elasticity so that we stretch and shrink them). So the crux of our
problem will lie in the form of the knot, not in how thick or long it is.
The reader familiar with topology will notice this discussion is just the
type of problem encountered in this branch of mathematics. To put it
plainly, knot theory, at the very least, may be considered to be a branch
of topology.

In fact, the knot in Figure 0.3 and the trivial knot depicted in
Figure 0.4 *are* equivalent. As our first "mathematical" exercise, we
leave it as a straightforward exercise for the reader to show by simple
manipulations that this indeed is the case. Emboldened by an early
success, we now ask ourselves, Is it true that the knots in Figures 0.1(b)
and 0.4 are equivalent? The answer is "no", however, this will not be
our second "mathematical" exercise since the proof is far from simple.
It could be said that knot theory owes its present development to this
very question being asked towards the end of the 19th century. In other
words, one aspect of knot theory is to provide the ways and means, to
a standard of mathematical rigour, to determine whether two knots are
equivalent.

Let us try our luck again: are Figures 0.1(b) and 0.2(a) equivalent?
The answer unfortunately is "no", and the proof again requires more
than just a modicum of intuition, so we do not leave it as an exercise.

It is difficult to say who first showed a mathematical interest in what we now call knot theory and when. However, in modern times it is known that the famous C.F. Gauss (1777–1855) had some interest in this field, but it was his student, Listing, who undertook research into knot theory and gave not insignificant influence to its later development. Originally, in honour of his accomplishments, the knot now known as the *figure 8 knot*, Figure 0.5, was called the *Listing knot*.

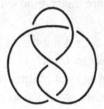

Figure 0.5

The American mathematician J.W. Alexander (1888–1971) was the first to show that knot theory is extremely important in the study of 3-dimensional topology. This was further underlined by the work of, amongst others, the German mathematician H. Seifert from the late 1920s to the 1930s. In addition, in Germany at this time there was already considerable activity in the study of the relationship between algebraic geometry and knot theory.

After the Second World War, in the 1950s, research into knot theory progressed at a great pace in the United States. Under the influence of this research, there was a great boom of research into knot theory in Japan, which has continued to the present day. In the 1970s knot theory was shown, among other things, to be connected to algebraic number theory, by virtue of the solution of Smith's conjecture concerning periodic mappings.

At the beginning of the 1980s, due to the discovery by V.F.R. Jones of his epochal knot invariant, knot theory moved from the realm of topology to mathematical physics. This was further underlined when it was shown that knot theory is closely related to the solvable models of statistical mechanics. As knot theory grows and develops, its boundaries continue to shift. Now, in addition, they overlap certain areas of mathematical biology and chemistry. To expand briefly on this, in biology certain types of DNA molecules have been experimentally seen to take the form of certain types of knots. In the chapters that follow, we shall introduce, hopefully in a fairly easy and understandable manner, the contributions of various mathematicians to knot theory and also the relation/application of knot theory to other fields.

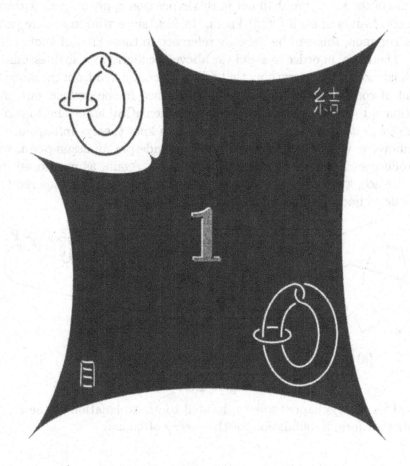

Fundamental Concepts of Knot Theory

A knot, succinctly, is an entwined circle. However, throughout this book we shall think of a knot as an entwined polygon in 3-dimensional space, Figure 1.0.1(a). The reason for this is that it allows us, with recourse to combinatorial topology,[1] to exclude *wild knots*. For an example of a wild knot, consider the knot in Figure 1.0.1(b). Close to the point P, in a sense we may take this to be a "limit" point, the knot starts to cluster together in a concertina fashion. Therefore, in the vicinity of such a point particular care needs to be taken with the nature of the knot. We shall not in this exposition apply or work within the constraints of such (wild) knots. In fact, since wild knots are not that common, this will be the only reference to these kind of knots.

Therefore, in order to avoid the above peculiarity, we shall assume, without exception, everything that follows is considered from the standpoint of combinatorial topology. As mentioned in the preface, our intention is to avoid as much as possible mathematical argot and to concentrate on the substance and application of knot theory. Infrequently, as above, it will be necessary, in order to underpin an assumption, to introduce such a piece of mathematical argot. Again, as mentioned in the preface, knowledge of such concepts will not usually be required to be able to understand what follows.

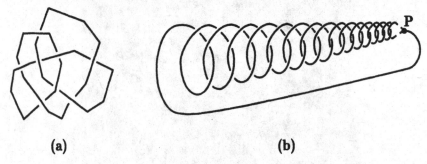

(a)   (b)

Figure 1.0.1

This (first) chapter will be devoted to an explanation of the concepts that form a foundation for the theory of knots.

§1 The elementary knot moves

If we consider a knot to be polygonal in form, then since it is possible to think of it as being composed of an immense (but still finite) number of edges, a knot is often depicted with *smooth* rather than polygonal arcs. As the reader can see by flicking through the book, we shall also follow this aesthetic criterion. However, mathematically, it *remains* a collection of polygonal lines.

Continuing towards a precise (mathematical) interpretation of a knot, it is readily óbvious that we can make alterations to the shape of a knot. For example, it is possible to replace an edge, AB, in space of a knot K by two new edges AC, CB. We can also perform the converse replacement. Such replacements are called *elementary knot moves*. We shall now precisely define the possible moves/replacements.

Definition 1.1.1. On a given knot K we may perform the following four operations.

(1) We may divide an edge, AB, in space of K into two edges, AC, CB, by placing a point C on the edge AB, Figure 1.1.1.

(1)′ [The converse of (1)] If AC and CB are two adjacent edges of K such that if C is erased AB becomes a straight line, then we may remove the point C, Figure 1.1.1.

(2) Suppose C is a point in space that does not lie on K. If the triangle ABC, formed by AB and C, does not intersect K, with the exception of the edge AB, then we may remove AB and add the two edges AC and CB, Figure 1.1.2.

(2)′ [The converse of (2)] If there exists in space a triangle ABC that contains two adjacent edges AC and CB of K, and this triangle does not intersect K, except at the edges AC and CB, then we may delete the two edges AC, CB and add the edge AB, Figure 1.1.2.

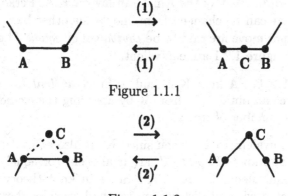

(1)
→
←
(1)′

A   B   **A C B**

Figure 1.1.1

(2)
→
←
(2)′

Figure 1.1.2

These four operations (1), (1)′, (2) and (2)′ are called the *elementary knot moves*. [However, since (1) and (1)′ are not "moves" in the usual understanding of this word, often only (2) and (2)′ are referred to as elementary knot moves. In this book we shall, by and large, also use this interpretation.]

§2 The equivalence of knots (I)

A knot is not perceptively changed if we apply only one elementary knot move. However, if we repeat the process at different places, several times, then the resultant knot *seems* to be a completely different knot. For example, let us look at the two knots K_1 and K_2 in Figure 1.2.1, which may be called *Perko's pair*.

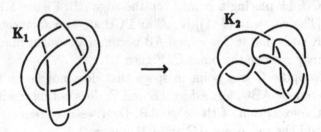

Figure 1.2.1

In appearance Perko's pair of knots looks completely different. In fact, for the better part of 100 years, nobody thought otherwise. However, it is possible to change the knot K_1 into the knot K_2 by performing the elementary knot moves a significant number of times. This was only shown in 1970 by the American lawyer K.A. Perko.

Knots that can be changed from one to the other by applying the elementary knot move are said to be *equivalent* or *equal*. Therefore, the two knots in Figure 1.2.1 are equivalent.

Definition 1.2.1. A knot K is said to be *equivalent* (or *equal*) to a knot K′ if we can obtain K′ from K by applying the elementary knot moves a finite number of times.

If K is equivalent to K′, then since K′ is also equivalent to K, we say that the two knots K *and* K′ are equivalent (or equal). We shall denote this equivalence by $K \approx K'$. Since in knot theory equivalent knots are treated without distinction, we shall consider them to be the *same*[2] knot.

The elementary knot move (2) allows us to replace an edge AB with the edges AC and CB. Since the points within the triangle ABC do not intersect with the knot itself, intuitively we may rephrase Definition 1.2.1 as follows: Two knots are equivalent if in space we can alter one continuously, without causing any self-intersections, until it becomes transformed into the other knot.

A knot has no starting point and no endpoint, i.e., it is a simple closed curve (to be precise a closed polygonal curve). Therefore, we can assign an orientation to the curve. As is the custom, we shall denote the orientation of a knot by an arrow on the curve. It is immediately obvious that any knot has two possible orientations, Figure 1.2.2.

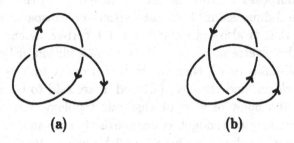

(a)   **(b)**

Figure 1.2.2

If two oriented knots K and K′ can be altered with respect to each other by means of *oriented* elementary knot moves, Figure 1.2.3, then we say K and K′ are equivalent with *orientation preserved* (or, for brevity, with *orientation*), and we write K ≅ K′.

Figure 1.2.3

Two knots that are equivalent without an orientation assigned are not *necessarily* equivalent (with orientation) when we assign an orientation to the knots. The two knots in Figure 1.2.2 are certainly equivalent without an orientation assigned; it is not, however, immediately obvious whether they are equivalent with orientation.

Exercise 1.2.1. Show that, in fact, the two knots in Figure 1.2.2 are equivalent with orientation.

§3 The equivalence of knots (II)

The elementary knot moves on a knot are "local" moves or transformations applied to only a small part of the knot itself. Instead of such local modifications, we can redefine an equivalence of knots in terms of "global" transformations/moves. These transformations move the whole space in which the knot exists. First, however, we need to explain briefly a few concepts that can be found in most textbooks on algebraic topology.

Let *f* be a map from a topological space X to a topological space

Y. For our purposes we can restrict our attention to the cases where X and Y are 3-dimensional Euclidean spaces or subspaces thereof. If f is a map that is also onto and has a 1-1 correspondence, then we may define the inverse map $f^{-1} : Y \to X$. In addition, if both f and f^{-1} are continuous maps, then the map f from X to Y is said to be a *homeomorphism*, and the spaces X and Y are said to be *homeomorphic*. From the point of view of algebraic topology, spaces that are homeomorphic may be thought as being exactly the same, i.e., without any distinctions. In the case where X and Y have orientations assigned to them, we say f is an *orientation-preserving homeomorphism* if the original orientation of Y *agrees* with the orientation on Y that is the effect of f on the orientation of X. Finally, a homeomorphism from X to itself, i.e., $X = Y$, is said to be an *auto-homeomorphism*.[3]

Example 1.3.1. Suppose that both X and Y are $\mathbf{R}^2$. Then the parallel translation along a line given by $(x, y) = (x + a, y + b)$; a rotation about some fixed point (for example, the origin) are examples of orientation-preserving (auto-)homeomorphisms, Figure 1.3.1(a).

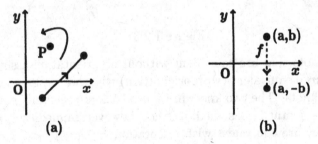

(a)   (b)

Figure 1.3.1

However, the mirror "image" with respect to the x-axis given by the homeomorphism $f(x, y) = (x, -y)$ is a not an orientation-preserving (auto)-homeomorphism, in fact, the orientation is reversed, Figure 1.3.1(b) (in this figure, the effect of the map f on the y-axis is to reverse its original orientation).

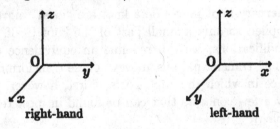

right-hand   left-hand

Figure 1.3.2

There is also a natural way of assigning an orientation to $\mathbf{R}^3$, which is done by means of the right-hand rule with regard to the xyz-axis, Figure 1.3.2.

Definition 1.3.1. We say that two knots K_1 and K_2 are equivalent, or K_1 is equivalent to K_2, if there exists an orientation-preserving homeomorphism of $\mathbf{R}^3$ that maps K_1 to K_2.

Although we now have two definitions of equivalence, Definitions 1.2.1 and 1.3.1, mathematically they are the same. A proof of this "equivalence" is given in Kawauchi [K*]. (*Nota bene*, in their proof it is not assumed that the necessary mapping is a PL-map.)

The (Euclidean) spaces in Example 1.3.1 are 2-dimensional, but it is not hard to see that if we move up a dimension, then a rotation about a fixed point (or fixed axis) and a parallel translation are examples of auto-homeomorphisms of $\mathbf{R}^3$, which preserve the orientation of $\mathbf{R}^3$. However, if we consider the mirror image, with regard to the xy-plane, given by $\varphi(x, y, z) = (x, y, -z)$, then this map reverses the orientation. For suppose that the xy-plane *is* a mirror, then we may think of φ as "reflecting" the point P in the mirror to a point P′, Figure 1.3.3. Similarly, the three axis with the right-hand rule are "reflected" to the three axis with the left-hand rule. So φ does not preserve orientation.

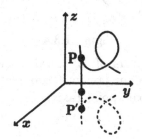

Figure 1.3.3

In general, we say a map φ is a *mirror image* (or a *symmetry*) with regard to a plane E if we can map an arbitrary point in $\mathbf{R}^3$ to its reflected point with regard to E. The image $\varphi(K)$, obtained from the effect of the mirror image φ on the knot K, is said to be the *mirror image* of K. If K has an orientation assigned, then we assign, in the obvious manner, an orientation to the mirror image of K from the orientation of K.

Consider, now, two knots, K_1 and K_2, both of which have orientations assigned. If we can map K_1 to K_2 by means of an orientation-preserving auto-homeomorphism of $\mathbf{R}^3$, that does not alter the orientation of *either* K_1 or K_2, then K_1 and K_2 are said to be equivalent with orientation.

The advantage of concentrating our attention on the definition of equivalence in Definition 1.3.1, rather than the one in Definition 1.2.1, which, let us recall, depends on elementary knot moves, is that from Definition 1.3.1 we may fairly immediately see, in a clear, intuitive way, a number of knot equivalences.

Example 1.3.2. Consider $\varphi(x,y,z) = (-x,-y,z)$, a 180° rotation about the z-axis, which is an orientation-preserving auto-homeomorphism, Figure 1.3.4. Since φ maps the oriented left-hand trefoil knot K to K′, these two knots are equivalent with orientation. K′ is the knot with the "reverse" orientation to K.

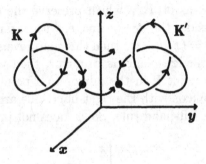

Figure 1.3.4

An auto-homeomorphism of $\mathbf{R}^3$ need not always move the whole of $\mathbf{R}^3$, as was the case in our previous examples: a parallel translation and a rotation about a fixed point (axis).

Example 1.3.3.

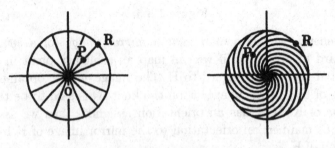

Figure 1.3.5

Let us fix a unit circle R, everything that lies "outside" this unit circle and the origin O. If we twist the "inside" of R about O, then this map is also an orientation-preserving auto-homeomorphism of $\mathbf{R}^2$, Figure 1.3.5.

This type of continuous "movement" is called an *isotopy*. In the above example a point P on the radius OR, rather like in a whirlpool, has been dragged by the isotopy towards the centre.

Similarly, fix the unit sphere in $\mathbf{R}^3$ and everything that lies outside the unit sphere. The map that twists the inside of the unit sphere about the x-axis is also an orientation-preserving auto-homeomorphism of $\mathbf{R}^3$. An example of a knot K' that has been obtained from K by such a twist is shown in Figure 1.3.6.

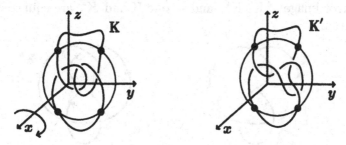

Figure 1.3.6

For the same reasons as given above, these two knots are equivalent (this map is called a *twist* of the ball, keeping the surface fixed).

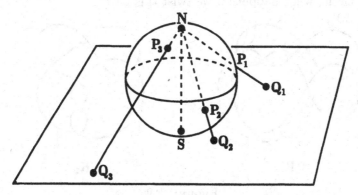

Figure 1.3.7

Figure 1.3.7 gives us a way of seeing the 1-1 correspondence between the 2-dimensional sphere, S^2, excluding the "North Pole" N, and the whole of the plane. So, if we add to the plane, $\mathbf{R}^2$, "a point at infinity," ∞, then $\mathbf{R}^2 \cup \infty$ and S^2 become homeomorphic (spaces).[4] Similarly, the 3-dimensional sphere, S^3, is often thought of as $\mathbf{R}^3$ with a point, ∞, added at infinity. On some occasions it will be more convenient for us to think of a knot lying in S^3 rather than $\mathbf{R}^3$. (The necessary adjustment to Definition 1.3.1 is to replace $\mathbf{R}^3$ by S^3. We assume, for obvious reasons, the knot K does not contain the point at infinity.)

Using this (re)definition, the following theorem can be seen to hold:

Theorem 1.3.1.

If two knots K_1 and K_2 that lie in S^3 are equivalent, then their complements $S^3 - K_1$ *and* $S^3 - K_2$ *are homeomorphic.*

Exercise 1.3.1. Show first that the knot K, shown in Figure 1.3.8, is the mirror image of K, K^*, and second K and K^* are equivalent with orientation.

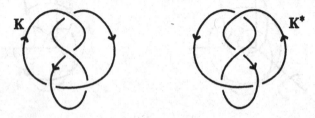

Figure 1.3.8

Exercise 1.3.2. Show that the two knots K_1 and K_2 in Figure 1.3.9 are equivalent. (Hint: Consider the part of the knot that lies within the dotted circle; what happens if we twist this part?)

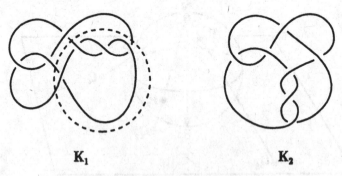

Figure 1.3.9

§4 Links

So far we have only looked at a rather specific, but interesting in its own right, set, namely, knots. In this section we shall look at a generalization of this set, i.e., what happens when we "link" a number of knots together.

Definition 1.4.1. A *link* is a finite, ordered collection of knots that do not intersect each other. Each knot K_i is said to be a *component* of the link.

Definition 1.4.2. Two links $L = \{K_1, K_2, \ldots, K_m\}$ and $L' = \{K'_1, K'_2, \ldots, K'_n\}$ are equivalent (or equal) if the following two conditions hold:

(1) $m = n$, that is, L and L' each have the same number of components;

(2)   We can change L into L' by performing the elementary knot moves a finite number of times. To be exact, using the elementary knot moves we can change K_1 to K'_1, K_2 to K'_2, $\ldots$, K_m to K'_n $(m = n)$. (We should emphasize that the triangle of a given elementary knot move does not intersect with any of the other components.)

We may replace (2) by the following (2)′ :

(2)′ There exists an auto-homeomorphism, φ, that preserves the orientation of $\mathbf{R}^3$ and maps $\varphi(K_1) = K'_1$, $\varphi(K_2) = K'_2, \ldots, \varphi(K_m) = K'_n$.

Strictly speaking, the equivalence of links should also be related to how we order the components. In general, however, such a stringent condition is not necessary, for we may suitably reorder the components. Usually, therefore, (2)′ is replaced by the following (2A):

(2A) There exists an auto-homeomorphism that preserves the orientation of $\mathbf{R}^3$ and maps the collection $K_1 \cup \ldots \cup K_m$ to the collection $K'_1 \cup \ldots K'_n$.

In this book we shall, on the whole make use of (2A) rather than (2)′. If each component of the link is oriented, then the definitions of equivalence are just an extension of the knot case.

Example 1.4.1. Since the two links L and L′ in Figure 1.4.1 are *exactly* the same, they are equivalent. However, if we change the order of the components of L, then condition (2) of Definition 1.4.2 is not satisfied and the links are not equivalent. But condition (2A) is satisfied; therefore, we shall consider them to be equivalent.

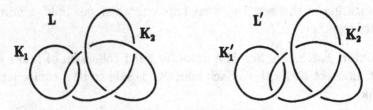

Figure 1.4.1

Now, let us assign an orientation to L and L', Figure 1.4.2. The addition of an orientation to the two links cause condition (2A) to no longer hold, and hence these oriented links are not equivalent. (To prove this we need to wait upon the definition of linking number in Chapter 4, Section 5.)

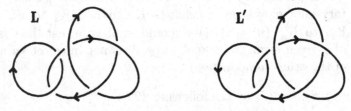

Figure 1.4.2

Due to this, we need to take some care when considering the problem of the equivalence of links, which is in direct contrast to the much less delicate problem of the equivalence of knots.

As might be expected, we may extend the concept of the trivial knot to the case of links, i.e., the trivial *n*-component link. In the extension to links, the relevant link consists of *n* disjoint trivial knots, Figure 1.4.3.

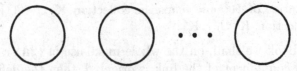

Figure 1.4.3

There is only *one* trivial *n*-component link for each *n* (orientating the trivial link has no bearing on this).

Propositions that hold with respect to knots and subsequently can be extended to links are numerous. Therefore, in what follows, if the examples, propositions, *et cetera* also hold for links, we will add "also holds in the case of links" at the end of the relevant statement or just "knots (or links)."

Exercise 1.4.1. Show that the two links in Figure 1.4.4 are equivalent. This link is called the *Whitehead link*.

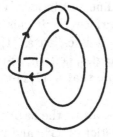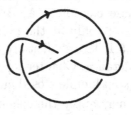

Figure 1.4.4

Exercise 1.4.2. Show that the two links in Figure 1.4.5 are equivalent. This link is called the *Borromean rings*.

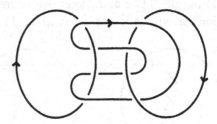

Figure 1.4.5

§5 Knot decomposition and the semi-group[5] of a knot

We may define a sum or product operation on the set that comprises all knots. If via such an operation the set becomes a group, then we might be able to apply group theoretical techniques to knot theory. Before we may try to investigate if this is possible, we must define such operations. Therefore, in this section we shall look into how we can obtain a single knot from two knots, called the sum (or connected sum) of these two knots. However, since a detailed explanation is a touch complicated, we shall delay a rigorous explanation until a bit later in the book. To get an insight into this approach we shall concentrate on the reverse operation, i.e., we shall *decompose* a knot (or link) into two *simpler* knots.

In this section it will be more convenient for us to think of the knot as lying in S^3.

Let us, now, consider a sphere, S, in S^3 (or $\mathbf{R}^3$) and the ball that is bounded by S, B^3, i.e., the 3-dimensional ball whose boundary is S. In the *interior* of B^3 take a simple curved line, α (in fact, a polygonal line) whose endpoints A, B are on the surface S. If this curve, α, intersects S only at the points A and B, it is called a $(1,1)$-tangle, Figure 1.5.1. [We shall study generalized tangles, i.e., (n,n)-tangles, in greater detail in Chapter 9.] We should note that a $(1,1)$-tangle may have disjoint simple closed curves.

We may apply the elementary knot moves to the segments of the knotted $(1,1)$-tangle, α that lie in the interior of B^3, and in so doing suppose that we change α, having fixed A and B, to the $(1,1)$-tangle shown in Figure 1.5.1(b). Such an α is called a *trivial* $(1,1)$-tangle. Figures 1.5.1(a),(b) are both examples of trivial $(1,1)$-tangles, while Figure 1.5.1(c) is an example of a $(1,1)$-tangle which is not a trivial $(1,1)$-tangle. [The ambitious reader might like to try to give an explanation for why it is not a trivial $(1,1)$-tangle.]

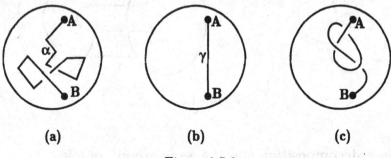

(a)   (b)   (c)

Figure 1.5.1

Suppose K is a knot (or link) in S^3. Further, let us suppose there exists a 2-dimensional sphere, Σ, that intersects (at right angles) K at *exactly* two points A and B. We may perceive Σ to fulfill the rôle of S described above. However, since K lies in S^3, K is divided by Σ into two $(1,1)$-tangles α and β, one of which lies within Σ and the other without, Figures 1.5.2(a) and 1.5.3(a). [Two $(1,1)$-tangles arise because Σ is the boundary of *two* 3-dimensional balls, one formed from Σ and its interior, and the other from Σ and its exterior. We may think of S^3 as being made up of two (3-dimensional) balls that have been glued together along their boundaries, namely, the 2-dimensional

sphere.[6] This gluing process is more easily visualized if we drop down
a dimension. For if we take two disks and glue them along their boundaries, in this case a circle, we obtain the 2-dimensional sphere.] Let us
now connect A to B by means of a simple polygonal line, s, that lies
on Σ. Then by joining s to α we obtain a knot K_1, and by joining
s to β, we obtain a knot K_2.

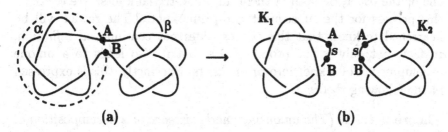

(a)   **(b)**

Figure 1.5.2

What we have shown is that a knot K can be decomposed into two
knots K_1 and K_2, Figure 1.5.2. The choice of s is arbitrary because
if we connect A to B by means of some other simple polygonal line that
lies on Σ, s', we shall once again decompose K into two knots say K_1'
and K_2'. It is reasonably straightforward to see that K_1 and K_1', and
K_2 and K_2', are equivalent (since we may apply the elementary knot
moves to s on Σ to change it into s'). If one of α or β, say, β,
is the trivial $(1,1)$-tangle then K_2' *is* the trivial knot. In such cases
K_1 and K_2 are not, strictly speaking, a "true" decomposition of K, see
Figure 1.5.3.

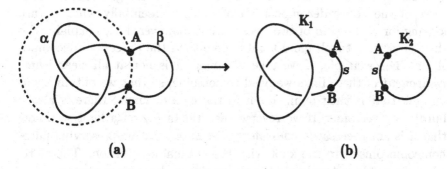

(a)   **(b)**

Figure 1.5.3

In fact, K and K_1 are equivalent, and so we do not think of K
as being decomposed into simpler knots. When a true (non-trivial)
decomposition cannot be found for K, we say that K is a *prime* knot.

In a sense this is equivalent to how we define a prime number, i.e., a natural number that cannot be decomposed into the product of two natural numbers, neither of which is 1.

From the above discussion, a knot K is either a prime knot or can be decomposed into at least two non-trivial knots. These non-trivial knots are either themselves prime knots, or we may, again, decompose one or the other, or both of them, into non-trivial knots. We continue this process for the subsequent non-prime knots. The reader will be heartened to know that this process does not continue *ad infinitum*. In fact, not only is the process finite, but it also leads to a unique decomposition of a knot into prime knots. Succinctly, this is expressed in the following theorem.

Theorem 1.5.1. *(The uniqueness and existence of a decomposition of knots)*

(1) *Any knot can be decomposed into a finite number of prime knots.*

(2) *This decomposition, excluding the order, is unique. That is to say, suppose we can decompose K in two ways:* $K_1, K_2, \ldots,$ K_m *and* $K'_1, K'_2, \ldots, K'_n$. *Then* $n = m$, *and, furthermore, if we suitably choose the subscript numbering of* $K_1, K_2, \ldots,$ K_m, *then* $K_1 \approx K'_1, K_2 \approx K'_2, \ldots, K_m \approx K'_m$.

A proof of this theorem can be found in Schubert [Sc1]. The above theorem also holds in the case of links.

Let us now think about the composition that brings about the converse of the above decomposition of knots. Essentially what we are looking for is the sum of two knots. First, however, let us explain why this "sum" is a touch more troublesome than the process of decomposition. For example, if we take two links it is not at all clear *which component* of these links we need to combine, so that we obtain a single link that is their sum. Even in the case of knots, there is also a hurdle to overcome. If we reverse the orientation on a knot we know that it is not necessarily equivalent, via an orientation-preserving auto-homeomorphism, to the knot with the original orientation. Therefore, when we combine two knots their orientations become important.

We shall show how to overcome the hurdle for two oriented knots. Suppose P is a point on an (oriented) knot K in S^3. We may think of P as the centre of a ball, $\mathbf{B}^3$, with a very small radius, Figure 1.5.4(a),(b), that possesses the following properties:

(1) K intersects (at right angles) exactly two points on the surface of boundary sphere of $\mathbf{B}^3$;

(2) In the interior of $\mathbf{B}^3$, the $(1,1)$-tangle, α, that is obtained from K is a trivial tangle.

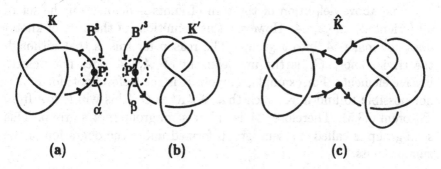

(a)   (b)   (c)

Figure 1.5.4

Similarly, to some other knot K′ in another 3-dimensional sphere S^3, we may choose a point P′ and, as above, obtain from K′ a trivial $(1,1)$-tangle, β, in some other ball $\mathbf{B}'^3$, Figure 1.5.4(b). We may in a natural way assign orientations, from K and K′ respectively, to the $(1,1)$-tangles α and β. Let $\widetilde{\mathbf{B}}^3$ be the ball that is obtained by removing from S^3 the points inside of $\mathbf{B}^3$. Similarly, let $\widetilde{\mathbf{B}}'^3$ be the ball that is obtained by removing from S^3 the points inside of $\mathbf{B}'^3$. The surface of each of these balls, i.e., $\widetilde{\mathbf{B}}^3$ and $\widetilde{\mathbf{B}}'^3$, is a (2-dimensional) sphere. If we now glue these two balls along this sphere, applying a homeomorphism that reverses throughout the orientation of the sphere of one of these balls, we obtain a 3-dimensional sphere, S^3. In gluing process the end (initial) point of α and the initial (end) point of β are joined. Therefore, in this 3-dimensional sphere, S^3, a *new, single, oriented* knot, $\widehat{K}$ is formed, Figure 1.5.4(c). By construction, the orientation of this $\widehat{K}$ will not contradict the original orientation of either K or K′.

The knot $\widehat{K}$ that is formed in the above process is said to be the *sum* of K and K′ (or the *connected sum*), and is denoted by K#K′. Moreover, this knot K#K′ is independent of the points P and P′ that were originally chosen. We can therefore say that K#K′ is uniquely determined by K and K′.

From the definition of the sum of knots, the next proposition follows readily.

Proposition 1.5.2.

The sum of two knots is commutative, i.e., $K_1 \# K_2 \cong K_2 \# K_1$. More concretely, $K_1 \# K_2$ and $K_2 \# K_1$ are equivalent with orientation. Also, the associative law holds, $K_1 \# (K_2 \# K_3) \cong (K_1 \# K_2) \# K_3$.

The above definition of the sum of knots is defined on the set of all (oriented) knots, $\mathcal{A}$. However, this definition (of the sum of knots) does not make $\mathcal{A}$ into a group. The reason for this is that although the trivial knot, O, is the unit element of $\mathcal{A}$, $\mathcal{A}$ does not possess inverse elements. For example, suppose K is the trefoil knot; for K it is not possible to find a K' such that $K \# K' \approx O$ (this will follow from Theorem 6.3.5). Therefore, $\mathcal{A}$ is only a semi-group, not a group. This semi-group is called the semi-group formed under the operation of the sum of knots.

Exercise 1.5.1. Show that the two knots in Figure 1.5.5 are equivalent.

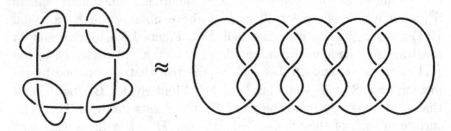

Figure 1.5.5

Exercise 1.5.2. Show that the two knots in Figure 1.5.6 are not equivalent. The knot in Figure 1.5.6(a) is called the *square knot*, while the one in Figure 1.5.6(b) is called the *granny knot*.

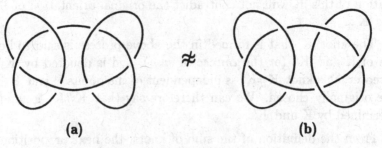

(a)   (b)

Figure 1.5.6

§6 The cobordism group of knots

We know from the above discussion that the set of all (oriented) knots is not a group under the most obvious operation, since inverse elements do not exist. Therefore, in order to ameliorate this situation and actually obtain a group, we may, for example, consider the following two possibilities:

(1) a change of the definition of the sum of two knots;

or

(2) a change of the definition of the equivalence of knots (or make it slightly weaker).

For example, the set of all integers under the action of multiplication is only a semi-group, while under the action of addition it becomes a group.

However, if we change the operation in such a manner for knots, then the algebraic structure becomes changed. So, it is *not* really an "improvement" of the semi-group obtained in the previous section. If, on the other hand, we slightly weaken the definition of *equality* (of knots), then perhaps the structure itself will not change considerably. At the beginning of the 1950s, J. Milnor introduced a new definition of equivalence called *corbordant*. With respect to this definition, the set of all knots does become a group under the action of the sum #. The group itself is called the *corbordism group* of knots, and the knots that become the unit element are called *slice knots*.

To explain the idea of a slice knot, consider S^3 as the boundary of a 4-dimensional ball $\mathbf{B}^4$ and take a knot K in S^3. The knot K is called a *slice knot* if it is the boundary of a disk, D, in $\mathbf{B}^4$ that does not have any singular "points." To make this precise, an interior point P of a disk D is *not* a singular point if we can always choose a neighbourhood U (homeomorphic to a 4-dimensional ball) of P in $\mathbf{B}^4$ in such a way that the intersection $\partial U \cap D$ is a trivial knot in ∂U, a 3-sphere.

For example, by joining an interior point Q of $\mathbf{B}^4$ with each point of a non-trivial knot K in S^3, the boundary of $\mathbf{B}^4$, we can construct a 2-dimensional surface F in $\mathbf{B}^4$ whose boundary is K. However, Q is a singular point of F, since for any neighbourhood V (homeomorphic to a 4-dimensional ball) of Q, $\partial V \cap F$ is equivalent to the original knot K in ∂V. The trivial knot, obviously, is a slice knot, but there are also many non-trivial knots that are slice knots.

Example 1.6.1. The square knot, shown in Figure 1.5.6(a), is a slice knot, but the trefoil knot and the figure 8 knot are not slice knots.

At the time of writing, no methods have been found that will detect exactly which knots are slice knots (see also Chapter 6, Section 4). It follows from the above "definition" that the study of the cobordism group is a problem that lies firmly in 4-dimensional topology, rather than within the realm of 3-dimensions. Finally, let us mention, without proof, a simple proposition concerning slice knots.

Proposition 1.6.1.
  Suppose K is an oriented knot, and $-K^*$ is the mirror image of K, with the orientation reversed. Then $K\# - K^*$ is always a slice knot.

Exercise 1.6.1. Show that the knot in Figure 1.6.1 is a slice knot.

Exercise 1.6.2. Show by repeated use of the elementary knot moves in $\mathbf{R}^4$ on any knot in $\mathbf{R}^4$ (to be precise a polygon in $\mathbf{R}^4$) that we can transform any knot into the trivial knot. For this reason, our definition of knot theory has no substance in 4-dimensions.

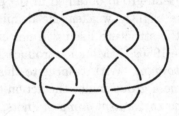

Figure 1.6.1

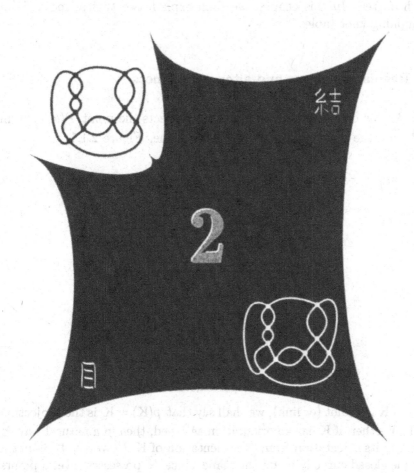

Knot Tables

Knot theory, in essence, began from the necessity to construct knot tables. Towards the end of the 19th century, several *mathematical* tables of knots were published independently by Little and Tait in British science journals. They managed to compile tables that in total consisted of around 800 knots, arranged in order from the simplest to the most "complicated." However, since these tables included, for example, the two knots in Figure 1.2.1 as "distinct" knots, these tables were subsequently found to be incomplete. However, considering that these lists were compiled around 100 years ago, they are accurate to a very high degree. In this chapter we shall explain two typical methods of compiling knot tables.

§1 Regular diagrams and alternating knots

Let us denote by p the map that projects the point $P(x, y, z)$ in $\mathbf{R}^3$ onto the point $\widehat{P}(x, y, 0)$ in the xy-plane, Figure 2.1.1.

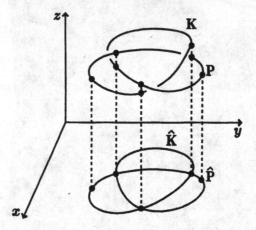

Figure 2.1.1

If K is a knot (or link), we shall say that $p(K) = \widehat{K}$ is the projection of K. Further, if K has an orientation assigned, then in a natural way $\widehat{K}$ inherits its orientation from the orientation of K. However, $\widehat{K}$ is *not* a simple closed curve lying on the plane, since $\widehat{K}$ possesses several points of intersection. But by performing several elementary knot moves on K – intuitively this is akin to slightly shifting K in space – we can impose the following conditions:

(1) $\widehat{K}$ has at most a finite number of points of intersection.

(2) If Q is a point of intersection of $\widehat{K}$, then the inverse image, $p^{-1}(Q) \cap K$, of Q in K has exactly two points. That is, Q is a double point of $\widehat{K}$, Figure 2.1.2(a); it cannot be a multiple point of the kind shown in Figure 2.1.2(b).

(3) A vertex of K (the knot considered now as a polygon) is never mapped onto a double point of $\widehat{K}$. In the two examples in Figure 2.1.2(c) and (d), a polygonal line projected from K comes into contact with a vertex point(s) of $\widehat{K}$, so both of these cases are not permissible.

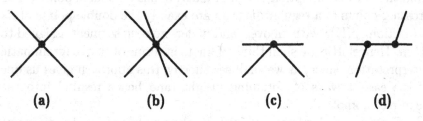

(a)   **(b)**   **(c)**   **(d)**

Figure 2.1.2

A projection $\widehat{K}$ that satisfies the above conditions is said to be a *regular projection*.

Throughout this book we will work almost exclusively with *regular* projections, *and* to simplify matters, we shall refer to them just as projections, we will draw a distinction only if some confusion might otherwise arise. However, even if we restrict ourselves to (regular) projections, there are still a considerable number of them; secondly, and at this juncture of quite some importance, is the ambiguity of the double points. At a double point of a projection, it is not clear whether the knot passes over or under itself. To remove this ambiguity, we slightly change the projection close to the double points, drawing the projection so that it *appears* to have been cut. Hopefully, this will give a *trompe l'oeil* effect of a continuous knot passing over and under itself. Such an altered projection is called a *regular diagram*, Figure 2.1.3(a),(b).

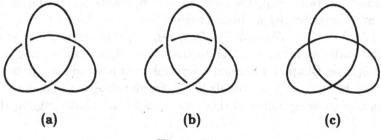

(a)   **(b)**   **(c)**

Figure 2.1.3

A regular diagram gives us a sense of how the knot may in fact lie in 3-dimensions, i.e., it allows us to depict the knot as a spatial diagram on the plane. Further, we can use the regular diagram to recover information lost in the projection, for example, Figure 2.1.3(c) is the projection of the two (non-equivalent) knots in Figure 2.1.3(a) and (b).

Therefore, we need to be a bit more precise with regard to the exact nature of a regular diagram and its crossing (double) points, since from the above description a regular diagram has no double points. The crossing points of a regular diagram are exactly the double points of its projection, $p(K)$, with an over- and under-crossing segment assigned to them. Henceforth, we shall think of knots in terms of this diagrammatic interpretation, since, as we shall see shortly, this approach gives us one of the easiest ways of obtaining insight (and hence results) into the nature of a knot.

For a particular knot (or link), K, the number of regular diagrams is innumerable. To be more exact, there is *only one* regular diagram of a knot, K, in $\mathbf{R}^3$. However, from our discussion in the previous chapter, the knot K and a knot K' obtained from K by applying the elementary knot moves are thought of as being the *same* knot. So, we can think of the regular diagram of K' as being a regular diagram for K. Hence, it follows that for K the number of regular diagrams is innumerable.

It is possible that a regular diagram may have crossing points of the type shown in Figure 2.1.4(a) and/or (b).

(a)   **(b)**

Figure 2.1.4

More generally, suppose two regular diagrams of two knots (or links) are connected by a single twisted band; see, for example, Figure 2.1.5(a) or (b). We can, in fact, remove this "central" crossing point by applying a twist, either to the left or right, to the knot, Figure 2.1.5(c) [in Chapter 1 Section 3 we explained how we can perform a twist that keeps the (2-)sphere fixed]. A regular diagram that does not possess any crossing points of this type is called a *reduced* regular diagram.

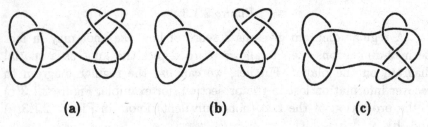

(a)   **(b)**   **(c)**

Figure 2.1.5

Let us now take an arbitrary point P on a regular diagram D of a knot K and move it once round D. If P, at the crossing points of D, is shown to, alternatively, move between a segment that passes over and a segment that passes under, then we say the regular diagram is an alternating (regular) diagram. Figures 1.5.5 and 1.5.6(b) are examples of alternating (regular) diagrams, while Figure 1.5.6(a) is an example of a non-alternating diagram. A knot that possesses (at least one) alternating diagram is called an *alternating knot*. These types of knots have great importance in knot theory, since many of their characteristics are known. (For a more detailed discussion, see Chapter 11, Section 5.) The trivial knot is also an alternating knot (we leave it as an exercise to explain why this is the case.) Many "simple" knots are alternating knots. Therefore, that is to say, in the nascent years of knot theory, all knots were thought to be alternating knots. The simplest non-alternating knot, in fact, is a knot with 8 crossing points shown in Figure 2.1.6. However, it is by no means trivial to prove that we can never find an alternating diagram for this knot. (For further details, see Chapter 7.)

Figure 2.1.6

Exercise 2.1.1. Show that a regular diagram for $K_1 \# K_2$ can be obtained by placing the regular diagrams of the oriented knots K_1 and K_2 side by side, and connecting them by means of two parallel segments, Figure 2.1.7.

Figure 2.1.7

Exercise 2.1.2. The definition of an alternating link follows directly from the definition of an alternating knot. Divide the knots and links that we have discussed so far into those with alternating diagrams and those that have non-alternating diagrams.

Exercise 2.1.3. Figures 1.5.6(a) and 2.1.7 are non-alternating diagrams for their respective knots. However, both of these knots are alternating knots. Show that they do possess alternating diagrams. (Hint: They have 6 and 7 crossings, respectively.)

Exercise 2.1.4. *(Taniyama)* Let K_1 and K_2 be alternating knots. Suppose that they have alternating diagrams with n_1 and n_2 crossing points, respectively. Show that the connected sum of K_1 and K_2 has an alternating diagram with exactly $n_1 + n_2$ crossing points.

§2 Knot tables

A table of reduced regular diagrams of knots may be thought of as a knot table. So, let us think how we may ascribe some sort of code/index system to these regular diagrams of knots. This aim is far from new. Gauss, probably as a recreation, devised one such code. Although other coding systems have been created, we shall describe, with a slight enhancement, the system due to Gauss [DT].

Suppose that a regular projection $\widehat{K}$ of a knot K has n crossing points, $\{P_1, P_2, \ldots, P_n\}$. Each crossing point P_i of $\widehat{K}$ is the projective image of exactly two points P_i' and P_i'' of K, Figure 2.2.1.

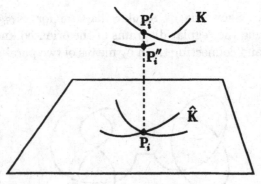

Figure 2.2.1

Now, starting with an arbitrary point P of K, move around K in a fixed direction (if K already has an orientation assigned, then follow K along this orientation). When we first arrive at a point P_i' or P_i'', assign the number 1 to this point. Moving on from the point P_i' or

P_i'', when we arrive at the next point P_j' or P_j'', assign the number 2 to this point (it is quite possible that we may assign the number 1 to P_i' and the number 2 to P_i''). In this way, we may assign to the $2n$ crossing points of K, $\{P_1', P_1'', P_2', P_2'', \ldots, P_n', P_n''\}$, the numbers 1 to $2n$, Figure 2.2.2.

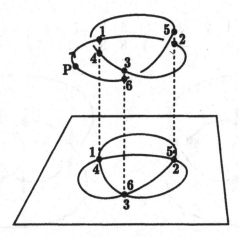

Figure 2.2.2

Due to this, we may assign two numbers to a point P_k of $\widehat{K}$, i.e., the projection of the points P_k' and P_k''. From these sets of pairs, (i, j_i) for each point P_k of $\widehat{K}$, we obtain a collection of $2n$ pairs of numbers,

$$(1, j_1), (2, j_2), \ldots, (2n, j_{2n}).$$

We shall rewrite these $2n$ pairs of numbers in the form of a permutation, i.e.,

$$N = \begin{pmatrix} 1 & 2 & 3 & \cdots & 2n \\ \pm j_1 & \pm j_2 & \pm j_3 & \cdots & \pm j_{2n} \end{pmatrix}.$$

The sign $+$ or $-$ in front of j_i obeys the following condition:

a "$+$" is assigned if the point of K that has the integer i
assigned to it is above the point that has the the integer   (2.2.1)
j_i assigned to it. If it is below, then we assign "$-$."

For example, the permutation corresponding to the knot in Figure 2.2.2 is

$$\begin{pmatrix} 1 & 2 & 3 & 4 & 5 & 6 \\ 4 & -5 & 6 & -1 & 2 & -3 \end{pmatrix}.$$

Example 2.2.1. The permutations that are obtained from the regular diagrams in Figure 2.2.3(a) and (b) are, respectively,

$$\begin{pmatrix} 1 & 2 & 3 & 4 & 5 & 6 & 7 & 8 \\ 4 & -7 & 6 & -1 & 8 & -3 & 2 & -5 \end{pmatrix}$$

and

$$\begin{pmatrix} 1 & 2 & 3 & 4 & 5 & 6 & 7 & 8 \\ 4 & 7 & -6 & -1 & 8 & 3 & -2 & -5 \end{pmatrix}.$$

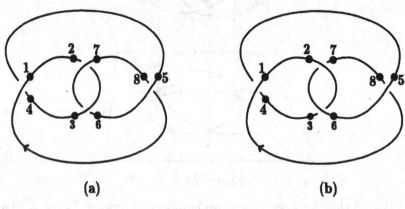

(a)   (b)

Figure 2.2.3

If we look closely at the permutations, then the following observations are almost immediate. Firstly, in the pair $(k, \pm j_k)$ one integer is always even and the other odd. (Why is this the case? Hint: Consider the Jordan curve theorem.[6]) Also, if the pair $(k, \pm j_k)$ is part of the permutation, then the inverse pair, $(j_k, \mp k)$, is also part of the permutation. Therefore, if we know the pairs that have an odd k, automatically the pairs that have an even k are also known. So, it is sufficient to consider only the permutations of odd-numbered pairs that originally comprised exactly half of the permutation. This now allows us to write down the following series, a row of n even numbers:

$$(\pm j_1, \pm j_3, \pm j_5, \ldots, \pm j_{2n-1}).$$

This series is the code assigned to (a regular diagram of) K.

Example 2.2.1. *(continued)* The code for the knot Figure 2.2.3(a) is $(4, 6, 8, 2)$, while for knot Figure 2.2.3(b) it is $(4, -6, 8, -2)$.

Exercise 2.2.1. Show that if all the signs in a given code agree, then it is a code of an alternating diagram; show that the converse also holds.

In the link case, choose one of its components and assign numbers
to this component in the manner described above. (As in the knot
case, if the component is oriented, then follow its orientation; otherwise,
the choice of direction in which we traverse the component is left to
the reader's discretion.) Next, choose another component and repeat
the above process, and continue until all the components have been
traversed and numbers assigned. In fact, if the starting points on each
component are suitably chosen, we may assign to each crossing point an
even number *and* an odd number. (Why is it possible to choose such
a starting point?) Hence, we may write down for each component a
row of even numbers in a similar manner as in the knot case. So, each
component will have a code assigned to it, and the sequence

$$(\pm j_1, \pm j_3, \ldots, \pm j_{2k-1} \mid \pm l_1, \pm l_3, \ldots, \pm l_{2m-1} \mid \ldots)$$

is the code of a link (diagram). The sign in front of the $j_i, l_i, \ldots$ is
determined in exactly the same way as in the case of knots. The symbol
$\mid$ between the row of j_i and l_i signifies that at this point the row of
even numbers for the first component comes to an end.

Example 2.2.2. The code for the (regular diagram of) Borromean
rings in Figure 2.2.4 is $(-6, -8 \mid -12, -10 \mid -2, -4)$.

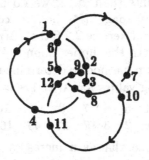

Figure 2.2.4

Exercise 2.2.2. Determine the codes for the square knot, the granny
knot, and the Whitehead link. Their regular diagrams were given in
Chapter 1.

We now have a method of assigning a code to a given knot K.
However, we immediately encounter a couple of problems. Firstly, the
code depends on the starting point, and, secondly, a knot, K, has an
abundance of regular diagrams. Hence each K has an abundance of
codes. Unfortunately, there is no known method to decide whether

or not two codes correspond to equivalent knots. However, we can determine whether or not a finite row of even integers is a code for some knot [DT].

Exercise 2.2.3. Suppose a sequence $(a_1, a_2, \ldots, a_n)$ is a code of a knot K. Show that the same sequence can be a code for the mirror image of K.

Exercise 2.2.4. *Find all* knots or links that have the following codes:
  (a) $(4, 8, -12, 2, 14, 16, -6, 10)$
  (b) $(6, 8, 22, 20, 4, -16, -26, -10, -24, -12, 2, -14, -18)$
  (c) $(6, 10, 2, -12 \mid 4, -8)$

Exercise 2.2.5. Show that there cannot exist a knot with the code $(8, 10, 2, 4, 6)$.

Exercise 2.2.6. Use the code of a knot to show that the number of knots and links that have regular diagrams with n crossing points is at most $2^n n!$

Let us now consider reduced regular diagrams with exactly n crossing points. Suppose that $\lambda(n)$ is the number of prime (unoriented) knots that have a regular diagram with n crossing points, but none with fewer crossing points than n. If we do not distinguish between a knot K and its mirror image K^*, i.e., we count them as the same knot, then we know from Exercise 2.2.6 that $\lambda(n) \leq 2^n n!$ In fact, $\lambda(n)$ is quite a bit smaller than this upper bound. However, at the time of writing, there is no known method to determine the exact value of $\lambda(n)$. At present, the following values for $\lambda(n)$ are known:

n	0	1	2	3	4	5	6	7	8	9	10	11	12	13
$\lambda(n)$	1	0	0	1	1	2	3	7	21	49	165	552	2176	9988

It is natural, of course, that as n increases, $\lambda(n)$ begins to increase rather rapidly. Actually, it was only a few years ago that it was *proven* that if n is large, then at the very least $\lambda(n)$ is bigger than n^2 [ES1]. Before this result was announced, basically all that could be said was that $\lambda(n) \geq 1$ for large n!

§3 Knot graphs

Let us first explain Tait's method for knots. Suppose that D is a regular diagram for a knot K and $\widehat{K}$ is a projection of K. We can think of $\hat{K}$ as a graph on the plane. (We shall explain the concept of a graph in a more detail in Chapter 14. For the present, the image we wish to use is that of the usual plane graph, i.e., composed of vertices and edges on the plane.) The vertices of the graph correspond to the crossing points of K.

In Figure 2.3.1 we have drawn a couple of plane graphs obtained from the two $\hat{K}$s in that figure.

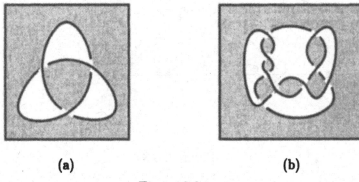

(a)   (b)

Figure 2.3.1

As can be seen from the above figures, $\hat{K}$ divides the plane into several domains. Starting with the outermost domain, we can colour the domains either black or white. By definition, we shall colour the outermost (unbounded) domain black. In fact, we can colour the domains so that neighbouring domains are *never* the same colour, i.e., on either side of an edge the colours never agree. (Why is it possible to colour domains in this manner?) Next, choose a point in each *white* domain; we shall call these points the centres of the white domains. If two white domains W and W' have the crossing points (of $\hat{K}$), $c_1, c_2, \ldots, c_l$, in common, then we connect the centres of W and W' by simple arcs that pass through $c_1, c_2, \ldots, c_l$ and lie in these two white domains (other than at the centres of W and W', these arcs do not intersect each other). In this way, we obtain from $\hat{K}$ a plane graph G. The vertices of G are the centres of the white domains.

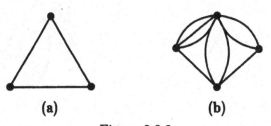

(a)   (b)

Figure 2.3.2

Example 2.3.1.   The plane graphs G of, respectively, Figure 2.3.2(a) and (b), are those obtained by the above method for the knots in Figure 2.3.1(a) and (b).

However, in order for the plane graph to embody some of the characteristics of the knot, we need to use the regular diagram rather than the projection. So, we need to consider the under- and over-crossing at a crossing. To this end, in Figure 2.3.3 is shown a way of assigning to each edge of G either the sign + or −.

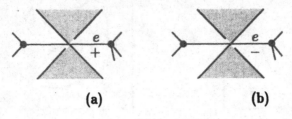

(a)   **(b)**

Figure 2.3.3

A + sign is assigned to an edge *e* if the domains are coloured in the manner of Figure 2.3.3(a), and a − sign if they are as in Figure 2.3.3(b). A signed plane graph that has been formed by means of the above process is said to be the *graph of* K. (To be precise, it is called the graph that is formed from the regular diagram D of a knot K.)

Example 2.3.2.   In Figure 2.3.4 we have drawn the signed plane graphs that correspond to the respective regular diagrams in that figure.

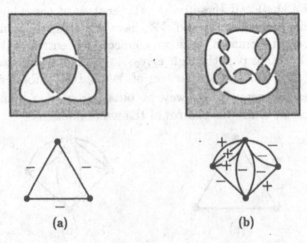

(a)   (b)

Figure 2.3.4

Conversely, we can construct from an arbitrary signed plane graph G a knot (or link) diagram; see Figure 2.3.5.

To construct the subsequent knot, first place a small "x" at the centre of each edge of G, Figure 2.3.5(b). From the endpoints of one of these of "x," draw four lines that follow along the edges of G until they reach the endpoints of a neighbouring "x." What should start to slowly appear if this process is carried out at each "x" is a projection of a knot, but with no information with regard to the nature of the crossing points, Figure 2.3.5(c). Now, we can colour the planar domains (obtained from the partition of the domain by the newly-formed projective diagram) either black or white using the same method to decide which colour to apply as discussed previously. We may ascertain, and hence draw in, the relevant crossing point information from the signs of the original graph. Obviously, the black and white colouring information disappears once the crossing point information is added, and hence we obtain the required knot diagram, Figure 2.3.5(d).

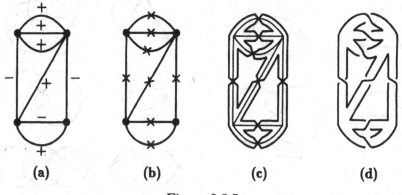

Figure 2.3.5

Therefore, for each signed plane graph there exists a corresponding (regular diagram of a) knot. However, it is not necessarily true that two different plane graphs give rise, by means of the above process, to two non-equivalent knots. At the time of writing, no method has yet been found to determine whether or not the two knots are equivalent.

The above approach was originally one of the methods used to construct a table of regular diagrams of all knots starting with graphs with a relatively small number of edges and then increasing the number of edges. In this manner, Tait and Little produced an almost complete table of regular diagrams of knots with up to 11 crossing points. In recent years this table has been amended and increased to include knots with up to 13 crossing points. [Appendix (I) is a complete table of all (prime) knots with up to 8 crossings.]

In Figure 2.3.6 we have placed in juxtaposition the *connected* plane graphs with up to 4 edges and their corresponding knots (and links). The number of edges is equal to the number of crossing points of the regular diagram of the knot. Since, for the sake of clarity, we have not assigned signs to the edges, these figures are not regular diagrams of the knots (or links) but rather their projections.

Graph   **Knot or Link**   **Graph**   **Knot or Link**

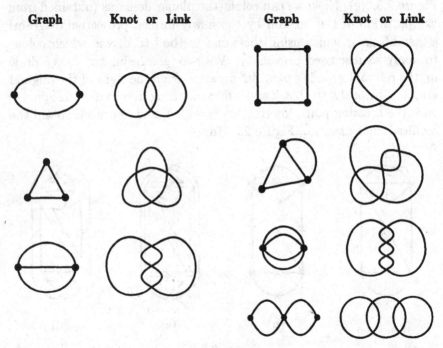

Figure 2.3.6

Exercise 2.3.1. Show that a regular diagram that is also an alternating diagram corresponds to a graph G with the same sign on all the edges. Moreover, show that these are the only possible kind of graphs.

Exercise 2.3.2. Why may we not think of an edge, *e*, as shown in Figure 2.3.7, to be an edge of a graph.

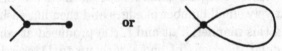

Figure 2.3.7

Exercise 2.3.3. List all the knots (and links) that correspond to connected plane graphs that have 5 and 6 positive edges. Moreover, determine which of these knots are equivalent.

To create a knot (or link) table, it is sufficient to create a table of prime knots (or links). A table of non-prime knots can be created directly from the table of prime knots. However, there is no known method that allows us to create a table of only prime knots. Moreover, at the time of writing, there is no known method to determine whether or not a given knot is prime (see also Chapter 3, Section 2).

Fundamental Problems of Knot Theory

The problems that arise when we study the theory of knots can essentially be divided into two types. On the one hand, there are those that we shall call *Global problems*, while, in contrast, there are those that we shall call *Local problems*.

Global problems concern themselves with how the *set of all* knots behaves. As the label implies, in contraposition, Local problems are concerned with the exact nature of *a given* knot. As to the question which is the more important, and hence we should concentrate our attention

tention on, the unhelpful answer is that it is impossible to say. In order to solve Global problems it is often necessary to find solutions to various Local problems. Conversely, the determination of Local problems may rely on how they fit within the Global problem.

In this chapter, we shall explain and give examples of these two types of problems. Problems in the theory of knots are not just limited to this bifurcation into Global and Local problems. However, in the past the above dichotomy has formed the axis around which knot theory has developed, and it is more than likely that this will substantially remain the case in the foreseeable future.

§1 Global problems

One of the typical classical Global problems is the classification problem.

(1)   The classification problem

The classification problem, at least in definition, is very straightforward, as the name suggests we would like to create a *complete* knot (or link) table. What exactly we mean by a complete table is one in which, firstly, no two knots are equivalent, and, secondly, a given arbitrary knot is equivalent to some knot in this table.

At the time of writing, a complete table in the above strict sense has been compiled only up to prime knots with 13 crossings. One future problem is to steadily expand this table. Another (sub-)problem that germinates directly from the original classification problem is to create a complete table for only certain specific types of knots, for example, for alternating knots. As we introduce other types of knots, this question of whether we classify them completely will always be in the vanguard of the questions that we will ask ourselves. In fact, in Chapters 7 and 9 we shall discuss two specific knot types that have been completely classified.

(2)   A fundamental conjecture

This conjecture can be immediately stated as follows:

> If $S^3 - K_1$ and $S^3 - K_2$, which are usually called complementary spaces, for two knots K_1 and K_2, respectively, are homeomorphic, then the knots are equivalent.

This conjecture can readily be seen to be the converse of Theorem 1.3.1.

In the late 1980s this conjecture was, in fact, proven by C. McA Gordon and J. Luecke [GL]. As a consequence of this result, the problem of knots in S^3 transforms itself from what we may call a *relative* problem which concerned itself with the shape of a knot in S^3, into an *absolute* problem, which now concerns itself with the study of the complementary spaces.

However, much to our dismay we cannot always transform a relative problem into an absolute problem. The counterexample that immediately comes to hand is that, in fact, the above fundamental conjecture is false in the case of links.

Example 3.1.1. Although the two links in Figure 3.1.1 are not equivalent, their complementary spaces are homeomorphic.[7]

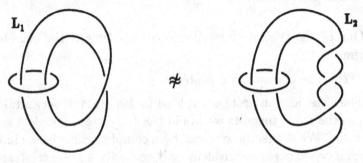

Figure 3.1.1

In general, results that hold for knots pass through fairly readily to hold for links as well. However, as the above example shows, we cannot take this for granted.

(3) Knot invariants

As a way of determining whether two knots are equivalent, the concept of the knot invariant plays a very important rôle. The types of knot invariants are not just limited to, say, numerical quantities. These knot invariants can also depend on commonly used mathematical tools, such as groups or rings.

Suppose that to each knot, K, we can assign a specific quantity $\rho(K)$. If for two equivalent knots the assigned quantities are always equal, then we call such a quantity, $\rho(K)$, a *knot invariant*. This concept of assigning some mathematical quantity to an object under investigation is not limited just to knot theory, it can be found in many branches of mathematics. Probably the simplest analogous example occurs in group theory. The number of elements in a group, called the order of the group, is a group invariant, since for isomorphic groups their respective orders are equal.

We know that if a knot K and another knot K' are equivalent, then it is possible to change K into K' by applying the elementary knot moves to K a finite number of times. Therefore, for a quantity $\rho(K)$ to be a knot invariant, $\rho(K)$ should not change as we apply the finite number of elementary knot moves to the knot K. It follows from this, for example, that the number of edges of a knot is not a knot invariant. The reason is that the operations defined in Definition 1.1.1(1) and (1)' either increase or decrease the number of edges. Similarly, if we consider the operations in (2) and (2)' of the same definition, then it also follows that the size of a knot is not a knot invariant.

A knot invariant, in general, is unidirectional, i.e.,

if two knots are equivalent $\overset{\text{then}}{\longrightarrow}$ their invariants are equal.

For many cases the reverse of this arrow does not hold. In contra-position, if two knot invariants are different then the knots themselves cannot be equivalent, and so a knot invariant gives us an extremely effective way to show whether two knots are *non*-equivalent. The history of knot theory may be said to be an account of how the various knot invariants were discovered and their subsequent application to various problems. To find such knot invariants is by definition a Global problem. On the other hand, to actually calculate many of these knot invariants, which we shall discuss in Chapter 4, is quite difficult. Further, to find a method to calculate these invariants is also a Global problem.

§2 Local problems

To illustrate and explain the idea of a Local problem, we shall give several examples.

(1) When are a knot K and its mirror image K* equivalent?

If K and K* are, in fact, equivalent, then we say that K is an *amphicheiral* knot (sometimes also referred to as an *achiral* knot). For example, since the right-hand trefoil knot [Figure 0.1(b)] and its mirror image, the left-hand trefoil knot (Figure 0.2), are not equivalent, the trefoil knot is *not* amphicheiral. On the other hand, however, the figure 8 knot is amphicheiral (cf. Exercise 1.3.1). Due to the extremely special nature of amphicheiral knots, there are, in relative terms, very few of them. This (Local) problem has over the years been quite extensively studied, and for particular types of knots many amphicheiral results have been proven. (For further details see Chapter 7 and Chapter 9, Section 3.)

(2) When is a given knot prime?

In the way described in Exercise 2.1.1 (Figure 2.1.7), a regular diagram of $K_1 \# K_2$, the connected sum of K_1 and K_2 may be constructed by placing the regular diagrams of K_1 and K_2 side by side and then connecting them by means of two parallel segments. Therefore, if a knot K can be decomposed into K_1 and K_2, then K has a regular diagram of the type shown in Figure 2.1.7. However, although theory predicts this in practice, since most regular diagrams of non-prime knots are usually not so nicely presented, we cannot deduce from the regular diagram whether a knot is prime.

Example 3.2.1. The regular diagram of the knot, K, shown in Figure 3.2.1(a) is not of the form of Figure 2.1.7, but K is not a prime knot.

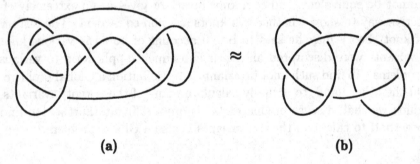

(a)   (b)

Figure 3.2.1

Recently, this (Local) problem has been completely resolved in the case of alternating knots (cf. Chapter 11, Section 5).

(3) When is a knot invertible?

We know that we can assign to a knot two different, opposite orientations. Let us denote one of these knots by K and the other, with the opposite orientation, $-K$. We would like to determine whether K and $-K$ are equivalent. When K and $-K$ are, in fact, equivalent, then K is said to be *invertible*. Knots with a relatively small number of crossing points are in general invertible. It follows from Example 1.3.2 that the left-hand trefoil knot is an example of an invertible knot.

That non-invertible knots do exist was first shown by H.F. Trotter in 1963.[8] The knot in Figure 3.2.2(a) was the example that was given by Trotter; following this discovery, many other non-invertible knots were soon found.

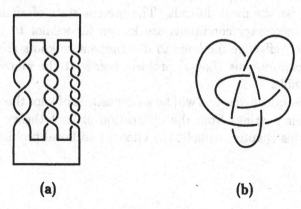

(a)   (b)

Figure 3.2.2

In contrast to 1963, it is now fair to say that almost all knots are non-invertible. We have drawn in Figure 3.2.2(b) the simplest non-invertible knot.

(4)   What is the period of a knot?

If we rotate the figure 8 knot, Figure 3.2.3(a), by an angle of π about the Oz-axis, the figure will rotate to its original form. So, this knot may be said to have period 2. The left-hand trefoil knot, Figure 3.2.3(b), if it is rotated by $\frac{2\pi}{3}$ about the Oz-axis, will also rotate to its original shape. In general, if we can rotate a knot by an angle $\frac{2\pi}{n}$ about a certain axis so that it rotates to its original shape, then we say that this knot has period n. In this case, the (Local) problem is to determine all the periods for a given knot. This problem has, also, been extensively studied and has been completely solved for particular types of knots (cf. Chapter 7).

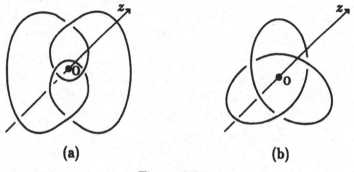

(a)   (b)

Figure 3.2.3

(5)   When is a knot a slice knot?

Of all the (Local) problems that we have so far discussed, this is probably by far the most difficult. The present state of affairs is that only several necessary conditions are known for a knot to be a slice knot. Further, effective methods to determine slice knots are also not known. Therefore, this (Local) problem seems at the moment to be quite intractable.

The subsequent chapters will be an exposition of knot theory, which will take their bearings from the bifurcation of knot theory problems outlined in this chapter, namely, the Global and Local problems.

Classical Knot Invariants

A knot (or link) invariant, by its very definition, as discussed in the previous chapter, does not change its value if we apply one of the elementary knot moves. As we have already seen, it is often useful to project the knot onto the plane, and then study the knot via its regular diagram. If we wish to pursue this line of thought, we must now ask ourselves what happens to, what is the effect on, the regular diagram if we perform a single elementary knot move on it? This question was studied by K. Reidemeister in the 1920s. In the course of time, many knot invariants were defined from Reidemeister's seminal work. In this chapter, in addition to discussing these types of knot invariants, we shall also look at knot invariants that follow naturally from what one might say is mathematical experience.

§1 The Reidemeister moves

A solitary elementary knot move, as might be expected, gives rise to various changes in the regular diagram. However, it is possible to restrict ourselves to just the four *moves* (strictly speaking, changes) shown in Figure 4.1.1 and their inverse moves, Theorem 4.1.1.

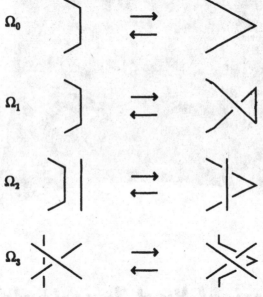

Figure 4.1.1

That these moves may, in fact, be made is reasonably straightforward to understand. For example, Ω_1 may be thought of as the move that corresponds to an elementary knot move on a regular diagram, which replaces AB by $AC \cup CB$, as shown in Figure 4.1.2.

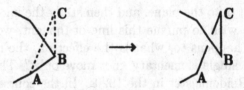

Figure 4.1.2

Exercise 4.1.1. Verify that, in fact, Ω_2 and Ω_3 are possible (i.e., they are a consequence of some finite sequence of elementary knot moves).

Example 4.1.1. The sequence of diagrams in Figure 4.1.3

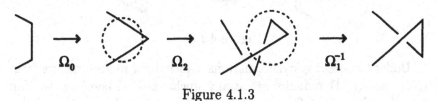

Figure 4.1.3

shows that the deformation

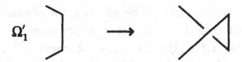

can be obtained as a sequence of Reidemeister moves.

Exercise 4.1.2. Show that the four deformations, and their inverses, in Figure 4.1.4 can be obtained as a sequence in the Reidemeister moves $\Omega_0, \Omega_1, \Omega_2, \Omega_3$ (*and* Ω_1') and their inverses.

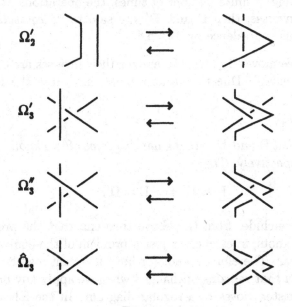

Figure 4.1.4

An obvious deformation like Ω_0 is one of the (plane) isotopic deformations defined in Chapter 1, Section 3, see Figures 1.3.5 and 4.1.5.

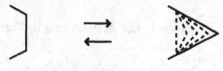

Figure 4.1.5

Under these isotopic deformations, which differ in their nature from Ω_1, Ω_2, or Ω_3, D remains essentially unchanged. Therefore, we can apply plane isotopic deformations quite freely to any place on the regular diagram, as long as D has a sufficient number of segments. To make sure that we have enough such segments, it may be necessary first to add a number of vertices to D. The addition (or elimination) of vertices on D corresponds exactly to the elementary knot move (1) [or (1)'] given in Definition 1.1.1. By the same reasoning, (1) and (1)' should not really be considered as moves (or changes), so we shall not classify them as an integral part of the moves. Keeping these remarks in mind, we can now define an equivalence between two regular diagrams D and D' of knots K and K'.

Definition 4.1.1. If we can change a regular diagram, D, to another D' by performing, a finite number of times, the operations $\Omega_1, \Omega_2, \Omega_3$ and/or their inverses, then D and D' are said to be *equivalent*. We shall denote this equivalence by $D \approx D'$.

These three moves $\Omega_1, \Omega_2, \Omega_3$ and/or their inverses are called the *Reidemeister moves*. Due to the above, we may state the following theorem.

Theorem 4.1.1.

Suppose that D *and* D' *are regular diagrams of two knots (or links)* K *and* K', *respectively. Then*

$$K \approx K' \Longleftrightarrow D \approx D'.$$

We may conclude, from the above theorem, that the problem of equivalence of knots, in essence, is just a problem of the equivalence of regular diagrams. Therefore, a knot (or link) invariant may be thought of as a quantity that remains unchanged when we apply any one of the above Reidemeister moves to a regular diagram. In the following, we shall often need to perform locally a finite number of times a composition

of Reidemeister moves (or plane isotopic deformations), for simplicity
we shall call such a composition an *R-move*.

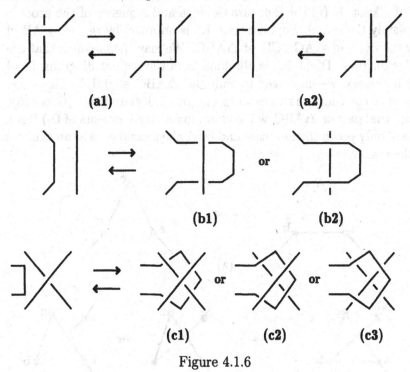

Figure 4.1.6

Lemma 4.1.2. *The moves shown in Figure 4.1.6 are R-moves.*

<u>Proof</u>

We prove two of the cases diagrammatically in Figure 4.1.7. The
other cases we will leave for the reader to prove along similar lines to
Figure 4.1.7.

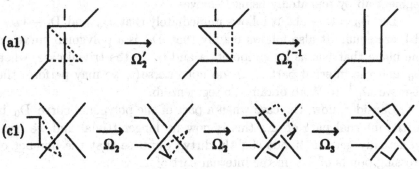

Figure 4.1.7

Proof of Theorem 4.1.1.

In order to simplify the notation we shall restrict the proof to the case of a knot, K (in the link case the idea and sequence of the proof is completely the same). Suppose that K' is obtained by replacing AB of K by the two edges $AC \cup CB$ of $\triangle ABC$. We may then assume that the regular diagram D' of K' is obtained from the regular diagram, D, of K (if necessary, we may need to shift the $\triangle ABC$ slightly). Therefore, we need to consider the two cases in Figure 4.1.8(a) and (b). (Generally, the internal part of $\triangle ABC$ will contain many line segments of D.) Here, we shall only prove the first case and leave the second case as an exercise for the reader.

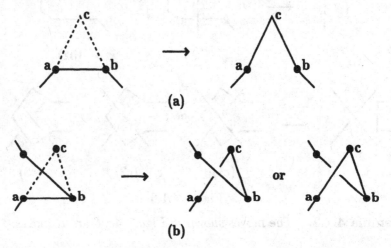

Figure 4.1.8

For the sake of clarity, we shall denote by $\triangle$ the projection, $\triangle abc$, of $\triangle ABC$. We shall now show that we can change $W = ac \cup cb$ to the segment ab by repeatedly using R-moves.

Let $D_0 = D - ab$. It follows immediately that D_0 and $D' - \{ac \cup cb\}$ are equal. It also follows readily that D_0 is a polygonal curve on the plane that has as its endpoints a and b. In the trivial case, when D_0 and the internal part of $\triangle$ do not intersect, we may perform the R-move Ω_0^{-1} to W to obtain the segment ab.

Consider, now, the case when a part of the polygonal curve D_0 is in the internal part of $\triangle$, this is now no longer trivial and we need to actually get our hands slightly dirty. Suppose that the number of crossing points of D_0 in the internal part of $\triangle$ is m.

First, by induction on these m crossing points, we will find, by applying R-moves to W, another simple polygonal curve W' on the plane. This (polygonal) curve W' will have as its endpoints the same a and b. In addition, inside the polygon formed from W' and the segment ab there will now be no crossing points of D_0. (Further, the only points where W' intersects with the segment ab is at the endpoints a and b.) To find such a W' we need the following definition.

Definition 4.1.2. A (not necessarily simple) polygonal curve α in D_0 that lies in the internal part of Δ (or, more generally, in the internal part of a polygon Σ on the plane) can be divided into two types:

α enters the internal part of Σ by passing over (or in the second type under) Σ at a point P and then exits the internal part of Σ by passing over (or in the second type under) at a point Q. Such an α (in relation to Σ) is called an *overlying* (or in the second type, *underlying*) polygonal curve, Figure 4.1.9.[9]

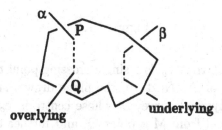

Figure 4.1.9

Lemma 4.1.3. *The polygonal curves of D_0 that are in the internal part of Δ are necessarily either overlying or underlying, (that is to say, there cannot exist a polygonal curve of D_0 that enters the internal part of Δ passing <u>over</u> Δ and the exits by passing <u>under</u> Δ). Further, at the intersection, in the internal parts of Δ, of an overlying polygonal curve α and an underlying polygonal curve β, α always passes above β.*

Proof

Let us consider $p^{-1}(\Delta)$, which, in fact, is a (infinitely long) triangular prism T in $\mathbf{R}^3$, Figure 4.1.10. This triangular prism is divided by $\triangle ABC$ into an upper prism T_1 and a lower prism T_2.

So, the question now follows: In which of these two prisms T_1 or T_2 does $p^{-1}(\alpha) \cap K$ lie, where $p^{-1}(\alpha)$ is the inverse image of the polygonal curve α of D_0 that lies in the internal part of Δ? Note that since K and the internal part of $\triangle ABC$ do not intersect, $p^{-1}(\alpha) \cap K$ cannot lie in both prisms. If $p^{-1}(\alpha) \cap K$ lies in T_1, then α is overlying, and if it lies in T_2, it is underlying. This proves the first part of the lemma. The latter half of the lemma also follows from the above.

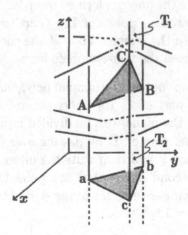

Figure 4.1.10

Suppose, now, that c_o is a single crossing point of D_0 that lies in the internal part of Δ. Next, on the plane draw a small circle (i.e., a simple closed polygonal curve) M whose centre is c_o. From W let us take a point P and from M a point Q, and further let us also choose a simple polygonal curve l, whose endpoints are P and Q. These are chosen in such a way that the following conditions are satisfied:

(1) P and Q are not vertices of either W or M, respectively;

(2) P and Q are not points of D_0 (i.e., P and Q are not crossing points of the regular diagrams D, D');

(3) l may intersect with D_0 at a finite number of points (at right angles); however, it cannot pass through a crossing point of D_0;

(4) l does not intersect with the segment ab.

We can always find points P and Q and a polygonal curve l, see Figure 4.1.11.

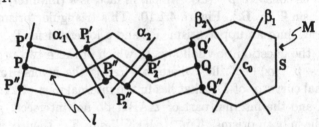

Figure 4.1.11

In the obvious manner make l slightly wider, i.e., we create a band $P'Q'Q''P''$. Our next move is to replace, by applying R-moves, the edge $P'P''$ by the other three edges of the band, $\gamma_1 = P'Q' \cup Q'Q'' \cup Q''P''$, Figure 4.1.11. To achieve this, firstly, if say, α_1, is the first overlying (underlying) polygonal curve that intersects l and D_0, then use the R-move (b1) [or (b2)] of Figure 4.1.6 to change $P'P''$ into $P'P_1'P_1''P''$, Figure 4.1.12(a) [or (b)].

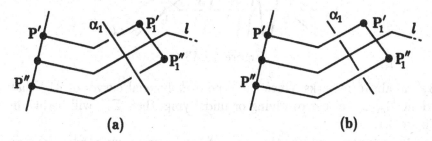

(a)   (b)

Figure 4.1.12

In a similar way, by repeating the process for the next intersection of l with a polygonal curve say, α_2, of D_0, we can replace $P_1'P_1''$ by $P_1'P_2'P_2''P_1''$. In this manner, we shall, finally, replace $P_k'P_k''$ by $P_k'Q'Q''P_k''$, and, so, we shall have completed the process of changing $P'P''$ into γ_1, where k is the number of points of intersection between D_0 and l. Next, replace $Q'Q''$, which is a part of M, by $M - Q'Q''$ (= $Q'SQ''$). Depending on the type of polygonal curve, β_1 and β_2, of D_0 that intersects with c_o, Figure 4.1.11, we shall need to perform the changes as described below. If both β_1 and β_2 are overlying, then apply the R-move (c1) shown in Lemma 4.1.2. If β_1 is overlying and β_2 is underlying, then apply (c2). Finally, if β_1 and β_2 are both underlying, then apply (c3).

The outcome of the above is that we replaced $P'P''$, a part of W, by another simple polygonal curve, $P'P_1' \ldots P_k'Q'SQ''P_k'' \ldots P_1''P''$, which we shall denote by W_1. This W_1 does not intersect with the segment ab. Therefore in the internal part of the polygon Σ_1 formed by W_1 and ab, the number of crossing points of D_0 has decreased by 1 to $m - 1$. Further, by the above operations, the polygonal curve of D_0 that lies in the internal part of Δ is divided into several polygonal curves. The types of the polygonal curves that remain in the internal part of Σ_1 are all the same as the original types of the polygonal curves.

We now repeat the above operations on the crossing points of D_0 that are in the internal part of Σ_1. At the end of this lengthy process, we shall reach a simple polygonal curve W_m that has as its endpoints a and b. In the polygon Σ_m, formed from ab and W_m, there are no crossing points.

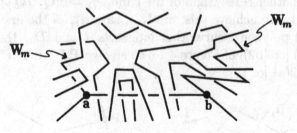

Figure 4.1.13

By the above remarks, since the type of polygonal curves of D_0 that are in Σ_m are either overlying or underlying, then Σ_m will be like in Figure 4.1.13.

In order to complete the proof of the theorem, we need to change W_m to the segment ab. This is done by repeatedly using R-moves given in Lemma 4.1.2. This final part we leave as an exercise to the reader.

■

In the next few sections, we shall explain several knot invariants that have played a substantial rôle in research into knots.

§2 The minimum number of crossing points

A regular diagram D of a knot (or link) K has at most a finite number of crossing points. However, this number c(D) is not a knot invariant. For example, the trivial knot has two regular diagrams D and D', which have a different number of crossing points, Figure 4.2.1.

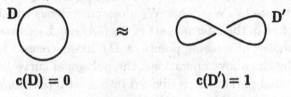

Figure 4.2.1

Consider, instead, all the regular diagrams of K, and let c(K) be the minimum number of crossing points of all the regular diagrams. This c(K) is a knot invariant.

Theorem 4.2.1.

$$c(K) = \min_{\mathcal{D}} c(D)$$

is a knot invariant, where $\mathcal{D}$ is the set of all regular diagrams, D, of K.

The above quantity is called the *minimum number of crossing points* of K. A regular diagram of K that has exactly $c(K)$ crossing points is said to be the *minimum regular diagram* of K. For example, if K is a trivial knot, then $c(K) = 0$.

<u>Proof of Theorem 4.2.1.</u>

Suppose that D_0 is the minimum regular diagram of K. Let K′ be a knot that is equivalent to K, and suppose that D_0' is its minimum regular diagram. Since we can think of D_0' as a regular diagram for K (K and K′ are equivalent), from the definition we have that $c(D_0) \leq c(D_0')$. However, since D_0 is a regular diagram of K′, it again follows from the definition that $c(D_0') \leq c(D_0)$. Hence, combining these two inequalities, we obtain $c(D_0) = c(D_0')$, i.e., $c(D_0)$ is the minimum number of crossing points for all knots equivalent to K. Consequently, it is a knot invariant.

∎

Exercise 4.2.1. Show that for $c(D) = 0, 1, 2$, the trivial knot is the only knot that possesses a regular diagram D with one of the above values.

Exercise 4.2.2. Show that the trefoil knot (either left-hand or right-hand), K, has $c(K) = 3$. Further, show that among all knots and links the trefoil knot is the only one with $c(K) = 3$.

Exercise 4.2.3. List all the knots and links with $c(K) = 2, 3, 4, 5$.

In general, there is no known method to determine $c(K)$. Recently, however, $c(K)$ has been completely determined in the case of alternating knots (or links), see Chapter 11, Section 5. For some *specific* types of non-alternating knots (or links), $c(K)$ has also been determined, see Chapter 7. However, the following conjecture has yet to be resolved:

Conjecture *Suppose that K_1 and K_2 are two arbitrary knots (or links), then*

$$c(K_1 \# K_2) = c(K_1) + c(K_2).$$

In the special case when both K_1 and K_2 are alternating knots (or
links), this conjecture has been shown to be true, see Chapter 11, Section 5.

§3 The bridge number

At each crossing point of a regular diagram, D, of a knot (or link)
K, let us remove (from D) a fairly small segment AB that passes over
the crossing point. The result of removing these segments is a collection
of disconnected (i.e., without any crossing points) polygonal curves, see
Figures 4.3.1(a) $\sim$ (c). We may think of the original regular diagram, D,
as the resulting diagram that occurs when we attach the segments AB,
..., (that pass over) to the endpoints of these disconnected polygonal
curves on the plane.

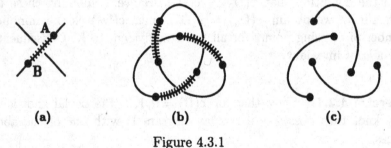

(a)   (b)   (c)

Figure 4.3.1

Since these segments AB pass above the segments on the plane,
these segments AB are called *bridges*. For a given D the number of
bridges is called the *bridge number*. To be more exact, let us introduce
the following definition:

Definition 4.3.1. Suppose that D is a regular diagram of a knot (or
link) K. If we can divide up D into $2n$ polygonal curves $\alpha_1, \alpha_2, \ldots, \alpha_n$
and $\beta_1, \beta_2, \ldots, \beta_n$, i.e.,

$$D = \alpha_1 \cup \alpha_2 \cup \ldots \cup \alpha_n \cup \beta_1 \cup \beta_2 \cup \ldots \cup \beta_n,$$

that satisfy the conditions given below, then the bridge number of D,
br(D), is said to be at most n.

(1) $\alpha_1, \alpha_2, \ldots, \alpha_n$ are mutually disjoint, simple polygonal curves.

(2) $\beta_1, \beta_2, \ldots, \beta_n$ are also mutually disjoint, simple curves.

(3) At the crossing points of D, $\alpha_1, \alpha_2, \ldots, \alpha_n$ are segments that pass *over* the crossing points. While at the crossing points of D, $\beta_1, \beta_2, \ldots, \beta_n$ are segments that pass *under* the crossing points.

If $\mathrm{br}(D) \leq n$ but $\mathrm{br}(D) \not\leq n - 1$, then we define $\mathrm{br}(D) = n$.

Example 4.3.1. The bridge number of the regular diagrams shown in Figure 4.3.2 are, respectively,

$$\mathrm{br}(D_1) = 3, \quad \mathrm{br}(D_2) = 2, \quad \mathrm{br}(D_3) = 2.$$

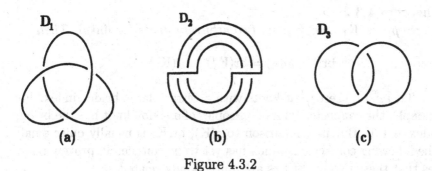

(a)   **(b)**   **(c)**

Figure 4.3.2

D_3 is another addition to our set of familiar and named knots and links; this link is called a *Hopf link*.

The bridge number of a regular diagram D is not a knot invariant for a knot K. There exist knots that have regular diagrams with different bridge numbers. In fact, Figure 4.3.2(a) and (b) are regular diagrams for the right-hand trefoil knot. (Show that these two diagrams are equivalent.) As in the previous section, if we consider all the regular diagrams for a given K, then the minimum bridge number of all these regular diagrams is an invariant for K.

Theorem 4.3.1.

For a knot (or link) K, $\mathrm{br}(K) = \min_{\mathcal{D}} \mathrm{br}(D)$ is an invariant for K, where $\mathcal{D}$ is the set of all regular diagrams of K. This quantity is called the bridge number (or the bridge index) of K.

Exercise 4.3.1. By considering the proof of Theorem 4.2.1, prove the above theorem.

Exercise 4.3.2. Show that if $\mathrm{br}(K) = 1$, then K is the trivial knot, and the trivial knot is the only knot with bridge number equal to 1.

Exercise 4.3.3. Show that if L is a n-component link then $\mathrm{br}(L) \geq n$. Unlike Exercise 4.3.2, if $\mathrm{br}(L) = n$ then L need not be the trivial link. For example, show that the Hopf link has $\mathrm{br}(L) = 2$.

In the specific case of $\mathrm{br}(K) = 2$, there are many knots with this bridge number, including the trefoil knot and the figure 8 knot. These knots, called for obvious reasons 2-bridge knots, have been extensively studied, to the point that they have been completely classified. In general, however, no method has yet been found to allow us to determine $\mathrm{br}(K)$ for an arbitrary knot K. But the following theorem has been proven in Schubert [Sc2].

Theorem 4.3.2.

Suppose K_1 and K_2 are two arbitrary knots (or links). Then

$$\mathrm{br}(K_1 \# K_2) = \mathrm{br}(K_1) + \mathrm{br}(K_2) - 1.$$

Therefore, there exist knots with arbitrary large bridge index. For example, the connected sum of n copies of a trefoil knot has the bridge index $n + 1$. But in comparison to $c(K)$, $\mathrm{br}(K)$ is usually quite small. The following conjecture, which has yet to be completely proven, signifies that these two quantities are quite closely related:

Conjecture. *If K is a knot, then*

$$c(K) \geq 3(\mathrm{br}(K) - 1),$$

where equality only holds when K is the trivial knot, the trefoil knot, or the (connected) sum of trefoil knots.

It is possible to calculate the bridge number in a different way to that described above. In order to redefine the bridge number so that we can calculate using the alternative method, we shall assume that the regular diagram of a knot (or link) K is a smooth (plane) curve D.

Figure 4.3.3

Denote by $\vec{v}(D)$ the number of local maxima of D, in relation to the direction of a certain (plane) vector $\vec{v}$, Figure 4.3.3.

As above, $\vec{v}(D)$ itself is not an invariant for K. However, if we consider all the regular diagrams for K, then the minimum value of all the number of local maxima, over all regular diagrams, is an invariant for K. This knot invariant is equal to br(K).

Theorem 4.3.3.

If we consider all the regular diagrams, D, of a knot K, then

$$\mathrm{br}(K) = \min_{D} \ \vec{v}(D).$$

Exercise 4.3.4.   Give a proof of the above theorem.

Exercise 4.3.5.   Determine the bridge number of the knots in Figure 4.3.4.

Exercise 4.3.6.   Find a regular diagram D of the square knot [Figure 1.5.6(a)] with br(D) = 3.

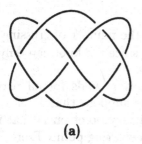   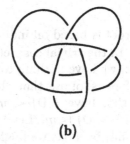

(a)   (b)

Figure 4.3.4

§4 The unknotting number

At one of the crossing points of a regular diagram, D, of a knot (or link) K exchange, locally, the over- and under-crossing segments. Since this type of alteration is not an elementary knot move, in general what we obtain is a regular diagram of some other knot.

Example 4.4.1.   In Figure 4.4.1(a), if we exchange the under- and over-crossing segments within the small circle, the subsequent regular diagram can readily be seen to be that of the trivial knot, Figure 4.4.1(b).

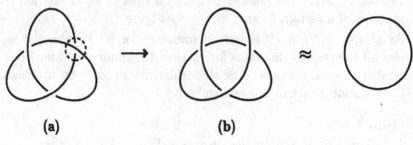

(a)   **(b)**

Figure 4.4.1

Proposition 4.4.1.

We can change a regular diagram, D, of an arbitrary knot (or link) to the regular diagram of the trivial knot (or link) by exchanging the over- and under-crossings segments at several crossing points of D (it may also be necessary to use the Reidemeister moves).

Due to Proposition 4.4.1, the above operation, which exchanges the over- and under-crossings segments at a crossing point, is called an *unknotting operation*.

<u>Proof</u>

The proof is based on induction on the number of crossing points, $c(D)$, of D. In the trivial case $c(D) = 0$, since the knot can only be the trivial knot, we have nothing to prove.

Therefore, suppose that the proposition holds for all regular diagrams D that have $c(D) < m$. Let us suppose that D is a regular diagram with $c(D) = m$. Let P, an arbitrary point on D that is not a crossing point, be what we might term a starting point. From P follow the knot around, naturally, in the direction of its fixed orientation.

If at a crossing point of D, we move along a part that passes over the crossing point, then do nothing just continue traversing the knot, Figure 4.4.2(a). However, if we arrive at a crossing point and then move along the part that passes under the crossing point, Figure 4.4.2(b), then at this crossing point perform an unknotting operation, Figure 4.4.2(c).

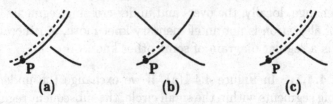

(a)   **(b)**   **(c)**

Figure 4.4.2

In this way, we will slowly create a regular diagram on which, starting from P, we shall always pass over the crossing points of the knot. If we continue traversing along D, repeating the above process, we shall eventually arrive at a crossing point A that we have already passed through, Figure 4.4.3(b). (If K is a link, then we may arrive back at our starting point P.)

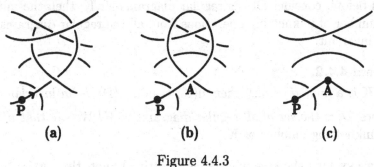

(a)   (b)   (c)

Figure 4.4.3

Once the above process has been finished, what will have been created is a loop that includes A, Figure 4.4.3(b). By applying Reidemeister moves, we may remove this loop. The new regular diagram, D', created by this process will have fewer crossing points than D. We may now apply to D' the induction hypothesis, in so doing we complete the proof.

∎

As in our previous discussions, we define the *unknotting number* of D as the minimum number of unknotting operations that are required to change D into the regular diagram of the trivial knot (or link). We will denote the unknotting number of D by u(D). As might be expected, u(D) is not an invariant of K.

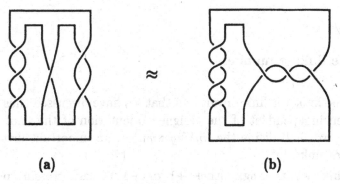

(a)   (b)

Figure 4.4.4

Exercise 4.4.1. In Figure 4.4.4 we have drawn two regular diagrams for a certain knot. Show that the regular diagram in Figure 4.4.4(a) requires only *one* unknotting operation to change it into a regular diagram of the trivial knot, while the regular diagram in Figure 4.4.4(b) requires *two* unknotting operations.

As before, consider all the regular diagrams for K, then the minimum number of unknotting operations from all the regular diagrams is a knot invariant.

Theorem 4.4.2.

If K is a knot (or link), then $u(K) = \min_{\mathcal{D}} u(D)$ *is an invariant of* K, *where* $\mathcal{D}$ *is the set of all regular diagrams of K. We say that* $u(K)$ *is the* unknotting number *of K.*

If we exclude the case when K is the trivial knot, then $u(K) \geq 1$. However, to, actually, determine $u(K)$ is a very hard problem. Even for very specific types of knots, there are virtually no methods, as yet, to determine $u(K)$.

Exercise 4.4.2. Show that the unknotting number of both knots in Figures 4.3.3 and 4.3.4(b) is 1, while the unknotting number of the knot in Figure 4.3.4(a) is at most 2.

Exercise 4.4.3. Show that the knot in Figure 2.1.6 has an unknotting number of at most 3.

Exercise 4.4.4. Show that it is possible to change an arbitrary regular diagram to an alternating regular diagram, which has the same projection, by performing the unknotting operation a finite number of times.

§5 The linking number

The knot (or link) invariants that we have discussed thus far have all been independent of the assigned orientation of the knot. In this section we shall define the *linking number*, an important invariant for *oriented* links.

First, let us assign either +1 or −1 to each crossing point of a regular diagram of an oriented knot or link.

Definition 4.5.1. At a crossing point, c, of an oriented regular diagram, as shown in Figure 4.5.1, we have two possible configurations. In case (a) we assign sign(c) = +1 to the crossing point, while in case (b) we assign sign(c) = −1. The crossing point in (a) is said to be positive, while that in (b) is said to be negative.

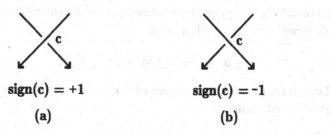

sign(c) = +1   **sign(c) = -1**

(a)   **(b)**

Figure 4.5.1

Suppose, now, that D is an oriented regular diagram of a 2-component link $L = \{K_1, K_2\}$. Further, suppose that the crossing points of D at which the projections of K_1 and K_2 intersect are $c_1, c_2, \ldots, c_m$. (We ignore the crossing points of the projections of K_1 and K_2, which are self-intersections of the knot component.)
Then

$$\frac{1}{2}\{\mathrm{sign}(c_1) + \mathrm{sign}(c_2) + \ldots + \mathrm{sign}(c_m)\}$$

is called the *linking number* of K_1 and K_2, which we will denote by $\mathrm{lk}(K_1, K_2)$.

Theorem 4.5.1.
  The linking number $\mathrm{lk}(K_1, K_2)$ *is an invariant of L.*

That is to say, if we consider another oriented regular diagram, D', of L, then the value of the linking number is the same as for D. Therefore, we shall call this number the *linking number* of L, and denote it by $\mathrm{lk}(L)$. Further, the linking number is independent of the order of K_1 and K_2, i.e., $\mathrm{lk}(K_1, K_2) = \mathrm{lk}(K_2, K_1)$.
  Before we give a proof of this theorem, we would like to consider a couple of examples and exercises.

Example 4.5.1. Let us calculate the linking number of the links L and L' in Figure 4.5.2(a) and (b), respectively.
  (a) We need only calculate the signs at the 4 crossing points $c_1, c_2, c_3,$ and c_4. Since the sign at each crossing point is −1, we obtain that $\mathrm{lk}(K_1, K_2) = -2$.

(b) Similarly, it is easy to show $\text{sign}(c_1) = \text{sign}(c_4) = +1$, while $\text{sign}(c_2) = \text{sign}(c_3) = -1$. Therefore, $\text{lk}(K_1', K_2') = 0$.

Exercise 4.5.1. Show that the linking number is always an integer. (Hint: Consider the Jordan curve theorem.[6])

Exercise 4.5.2. Suppose that we reverse the orientation of K_2, which we will denote by $-K_2$. Show that

$$\text{lk}(K_1, -K_2) = -\text{lk}(K_1, K_2).$$

Therefore, the linking number of L is an invariant that depends on the given orientation.

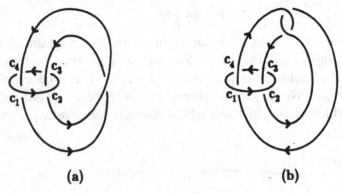

(a)   (b)

Figure 4.5.2

Proof of Theorem 4.5.1

Suppose that D' is another regular diagram of L. From our discussions thus far, we know that we may obtain D' by performing, if necessary several times, the Reidemeister moves $\Omega_1^{\pm}, \Omega_2^{\pm}$, and $\Omega_3^{\pm}$, Figure 4.1.1. Therefore, in order to prove this theorem it is sufficient to show that the value of the linking number remains unchanged after each of $\Omega_1^{\pm}, \Omega_2^{\pm}$, and $\Omega_3^{\pm}$ is performed on D. We shall only prove the theorem for the case when Ω_i $(i = 1, 2, 3)$ is applied and leave the remaining cases as exercises for the reader.

(1) At the crossing points of D at which we intend to apply Ω_1, every section (edge) of such a crossing point belongs to the same component. Therefore, applying a Ω_1 does not affect the calculation of the linking number.

(2) An application of Ω_2 on D only has an effect on the linking number if A and B, see Figure 4.5.3(a), belong to different components.

Since A and B can be assigned two different orientations, Figure 4.5.3(b) and (c), it is necessary to consider these cases separately. However, in both cases, since the newly created crossing points c_1 and c_2 have opposite signs, we have $\text{sign}(c_1) + \text{sign}(c_2) = 0$; again, the linking number is unaffected.

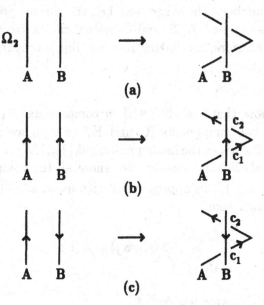

$$\Omega_2$$

A   B

(a)

A   B

A   B

(b)

A   B

A   B

(c)

A   B

Figure 4.5.3

(3)   Finally, let us consider the effect of Ω_3 on D, see Figure 4.5.4, i.e., the effect on the signs of c_1', c_2', c_3' and c_1, c_2, c_3, the crossing points that are affected by Ω_3.

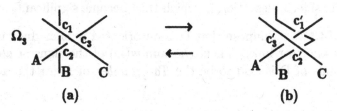

$$\Omega_3$$

(a)   **(b)**

Figure 4.5.4

Irrespective of how we assign the orientation on A, B, and C, the following equations will always hold:

$$\text{sign}(c_1) = \text{sign}(c_2'), \quad \text{sign}(c_2) = \text{sign}(c_1'),$$
$$\text{sign}(c_3) = \text{sign}(c_3'). \tag{4.5.1}$$

If A, B, and C all belong to the same component, then, as before, the linking number is unaffected. So suppose, first, that A belongs to a *different* component than B and C. Then the only parts that have an effect on the linking number is the sum $\text{sign}(c_2) + \text{sign}(c_3)$ in Figure 4.5.4(a) and $\text{sign}(c_1') + \text{sign}(c_3')$ in Figure 4.5.4(b). Due to (4.5.1), these two are, in fact, equal, and therefore this does not cause any change to the linking number. The other case, i.e., the various possibilities for the components to which A, B, and C belong, can be treated in a similar manner. Therefore, the linking number, lk(L), remains unchanged when we apply Ω_3.

∎

Suppose, now, that L is a link with n components, $K_1, K_2, \ldots, K_n$. With regard to two components K_i and K_j $(i < j)$, we may define as an extension of the above the linking number lk(K_i, K_j), $1 \leq i < j \leq n$. (To calculate this linking number, we ignore all the components of L except K_i and K_j.) This approach will give us, in all, $\frac{n(n-1)}{2}$ linking numbers, and their sum,

$$\sum_{1 \leq i < j \leq n} \text{lk}(K_i, K_j) = \text{lk}(L),$$

is called the *total linking number* of L.

Exercise 4.5.3. Show that, in fact, the total linking number of L is an invariant of L.

So far we have ignored the crossing points of K_1 itself or K_2. We shall now consider a definition in which they become significant.

Definition 4.5.2. Suppose that D is an oriented regular diagram of an oriented knot (or link). Then, the sum w(D) of the signs of *all* the crossing points of D is said to be the *Tait number* of D (or the *writhe* of D).

The Tait number of D, w(D), is itself not an invariant of a knot (or link). (Why is this the case?) As the name suggests, w(D) was first considered by P.G. Tait at around the turn of the 20th century. He thought that if D and D' are two *minimum* (in terms of the number of crossing points) regular diagrams of a knot, K, then $w(D) = w(D')$. It was believed for a long time that indeed this was the case. For this reason the diagrams in Figure 1.2.1 were thought to represent different knots. In fact, they are two different regular diagrams of the same knot!

Exercise 4.5.4. Calculate the Tait number of the two knot diagrams in Figure 1.2.1.

Exercise 4.5.5. Calculate the linking number of Figure 4.5.5(a) and (b).

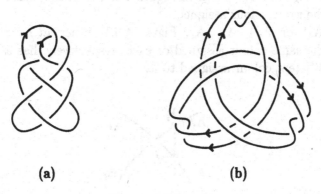

(a)   (b)

Figure 4.5.5

Exercise 4.5.6. Let L^* be the mirror image of an oriented link L. Show that $lk(L^*) = -lk(L)$.

Figure 4.5.6

Exercise 4.5.7. Show that the (unoriented) link in Figure 4.5.6 is not amphicheiral. [Hint: Give various orientations to each component of L and compare $lk(L)$ and $lk(L^*)$, where L^* is the mirror image of L.]

§6 The colouring number of a knot

In this final section of this chapter we would like to describe and define an invariant that is called the colouring number of a knot.

Suppose that the projection $\widehat{K}$ of a knot (or link) K has n crossing points $P_1, P_2, \ldots, P_n$. Since each P_i is the projection of the points P_i' and P_i'' of K, Figure 2.2.1, we can, by means of these points, divide

K into $2n$ segments (or polygonal curves) $A_1, A_2, \ldots, A_{2n}$. To each of these segments we may assign one of three colours red, blue, or yellow in such a way that the following two conditions are satisfied:

(1) If A_k and A_l are as in Figure 4.6.1, then they have the same colour assigned.

(2) A_k (or A_l), A_r, A_s, Figure 4.6.1, either *all* have the same colour assigned or each, respectively, has a different colour assigned to it. (14.0.1)

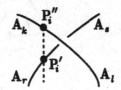

Figure 4.6.1

A regular diagram, D, of K, which *can* have the three colours assigned throughout the diagram in the above fashion, is said to be *3-colourable*.

Example 4.6.1. (a) The regular diagram, D, of the trefoil knot, K, as drawn in Figure 4.6.2, is 3-colourable. For example, we may assign red to the segment AB, blue to the segment BC, and yellow to the segment CA.

(b) The regular diagram of the figure 8 knot, Figure 0.5, is not 3-colourable.

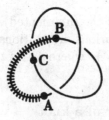

Figure 4.6.2

Proposition 4.6.1.

If there exists a regular diagram D of a knot (or link) K that is 3-colourable, then every regular diagram, D', of K is 3-colourable. Such a knot, K, is said to be 3-colourable.

<u>Proof</u>

Suppose that D is a 3-colourable regular diagram of K. If D' is another regular diagram of K, then, as before, we can change D into D' by applying, possibly several times, the Reidemeister moves $\Omega_1, \Omega_2, \Omega_3$ and their inverses. Therefore, to prove this proposition it is sufficient to show that each of the regular diagrams obtained after we have performed one of the Reidemeister moves $\Omega_1^\pm, \Omega_2^\pm$, and $\Omega_3^\pm$ is 3-colourable. (Note that the conditions for 3-colourability do not depend on the orientation of the knot, so we will ignore orientations.)

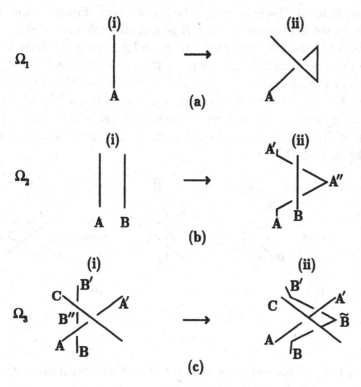

Figure 4.6.3

(1) In the case of a Ω_1-move, Figure 4.6.3(a), we may colour each segment of Figure 4.6.3(a)(ii) with the same colour as A. The same is true in the case of Ω_1^{-1}, which is left as an exercise for the reader.

(2) In the case of a Ω_2-move, Figure 4.6.3(b), if A and B are the same colour, then A' and A" may also be assigned the same colour. However, if A and B have different colours assigned to them, then assign to A' the same colour as A, and to A" assign the colour that neither A or B is coloured with, i.e., the third colour. The same arguments hold

in the case of Ω_2^{-1}, which also is left as an exercise for the reader.

(3) In the case of a Ω_3-move, Figure 4.6.3(c), if before we apply this move, Figure 4.6.3(c)(i), A, B, C, ..., all have the same colour, then after the Ω_3-move is applied, Figure 4.6.3(c)(ii), we may assign to them this same colour. In the general case, for clarity let us suppose that the three colours are denoted by α, β, γ. Now, let us consider the case shown in Figure 4.6.4(a)(i), where A and B are both assigned the colour α and C is assigned the colour β. Since A', B'', and B' are assigned the colours γ, α, and γ, respectively, in Figure 4.6.4(a)(ii) we may assign to $\widetilde{B}$ the colour γ. The cases where, firstly, A and C are both assigned the colour α and B is assigned the colour β and, secondly, B and C are assigned the colour α and A is assigned the colour β, can be dealt with in the same way as the preceding case. Finally, let us consider the case where A, B, and C *each* have a different colour assigned to them, α, β, and γ, say, Figure 4.6.4(b). Since B'', A', and B' have the colours γ, β, and γ, respectively, assigned to them, we may assign to $\widetilde{B}$, Figure 4.6.4(b)(ii), the colour α. The same argument holds in the case of Ω_3^{-1}.

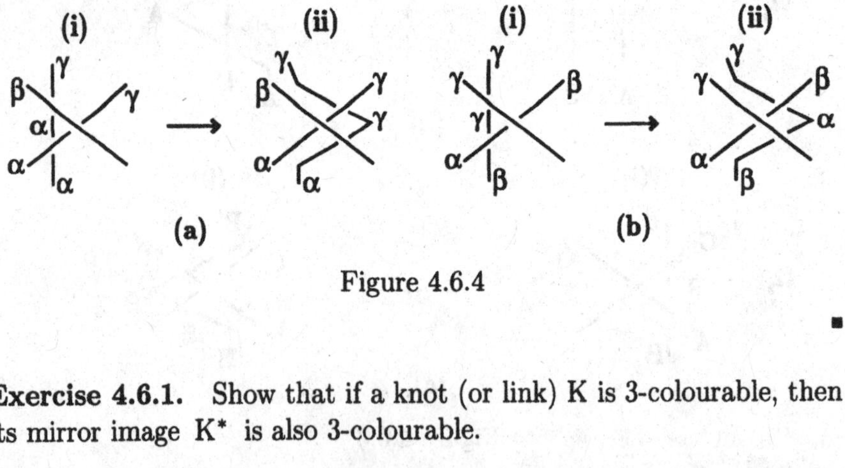

Figure 4.6.4

Exercise 4.6.1. Show that if a knot (or link) K is 3-colourable, then its mirror image K* is also 3-colourable.

In this section, so far, we have only considered *3*-colourable knots (or links). The reader may have already wondered: Is it possible to have p-colourable knots (or links) where p is a prime number.

As above, consider $A_1, A_2, \ldots, A_{2n}$, which we created previously, and assign to each segment (or polygonal line) A_i an integer λ_i that takes its value from the set of consecutive integers 0 to $p-1$, inclusive. These λ_i are assigned so that the following conditions are satisfied, with regard to Figure 4.6.1:

(1) $\lambda_k = \lambda_l$
(2) $\lambda_r + \lambda_s \equiv \lambda_k + \lambda_l \pmod{p}$.[10]

It is always possible to find λ_i for which the above conditions hold.
For example, in the extreme case we may assign $\lambda_1 = \ldots = \lambda_{2n} = 0$. In
the particular case when all the segments have the same colour assigned
to them, the colouring is said to be a *trivial colouring* of a regular
diagram, D. A regular diagram, D, is said to be *p*-colourable, if, at the
very least, two segments are assigned two different integers from the set
0 to $p-1$. The following proposition may be proven along similar lines
to Proposition 4.6.1:

Proposition 4.6.2.
   *If a knot (or link) K has at least one p-colourable regular diagram,
then every regular diagram is p-colourable.*

A given knot (or link), K, may be p-colourable with regard to several different *p*s. Therefore, the different number of colours with which
K may be coloured is an invariant of K. This invariant is the *colouring
number set* of K. It is possible to determine the colouring number set
for K [F2].

Exercise 4.6.2.   Show that a knot cannot be non-trivially 2-coloured,
but a link with $n\, (\geq 2)$ components can always be non-trivially
2-coloured.

Exercise 4.6.3.   Show that the figure 8 knot is 5-colourable but not
3-colourable.

Exercise 4.6.4.   The knots in Figure 4.6.5 are 3-colourable. Show that
this is the case. (Note: The method of colouring them is not unique.)

Figure 4.6.5

Exercise 4.6.5.   For the case $p = 3$, show that the definition of
p-colourable agrees with our original definition of 3-colourable.

Exercise 4.6.6. (1) Find two different 3-colourings of the diagram of the right-hand trefoil knot [Figure 0.1(c)] and show that one of the 3-colourings can be obtained from the other by simply permuting the 3 colours. (Such two colourings are called *equivalent*.)

(2) Find two non-equivalent 3-colourings of the (regular diagram of the) square knot, Figure 1.5.6(a). How many non-equivalent 3-colourings does this knot (diagram) have?

Seifert Matrices

In any science, in any discipline there are moments that can be called turning points – they reinvigorate and deepen the understanding of the subject at hand. What exactly is a turning point, even among friends, is usually contested and debated feverishly. Knot theory also has many turning points; however, there are two that are beyond debate: the Alexander polynomial and the Jones polynomial.

The Alexander polynomial, discovered by J.W. Alexander in 1928, has become one of the cornerstones of knot theory. Although the polynomial carries the epithet Alexander, Reidemeister at essentially the same time announced something that he called the L-polynomial. These two polynomials can be shown to be the same, even though Reidemeister's approach is independent of the work of Alexander. (The "L" in the L-polynomial is an abbreviation of Laurent, since in this polynomial some of the terms may have negative exponents, for example, t^{-2}.) We will deal with the second turning point, or more accurately, the Jones "revolution," in Chapter 11.

The study of the Alexander polynomial and determination of its precise context in knot theory has been extensively carried out. It has been found that the Alexander polynomial is very closely connected with the topological properties of the knot, and this has had a very profound impact on the theory of knots. Also of great significance is that there are various methods by which we may calculate the Alexander polynomial. In the mechanics of one of these methods, a concept called the *Seifert matrix* is brought to light. This concept of the Seifert matrix is itself extremely interesting and is, in fact, also one of the cornerstones of classical knot theory. This chapter is devoted to an explanation of the Seifert matrix and its properties. This will allow us, in the next chapter, to define the Alexander polynomial via Seifert matrix.

§1 The Seifert surface

Let us begin with the following theorem due to L. Pontrjagin and F. Frankl:

Theorem 5.1.1.

Given an arbitrary oriented knot (or link) K, then there exists in $\mathbf{R}^3$ *an orientable, connected surface, F, that has as its boundary K. (That is to say, there exists an orientable connected surface that spans K.)*

The above theorem was proven in 1930; here, however, we shall give a very neat proof due to Seifert.

<u>Proof</u>

Suppose that K is an oriented knot (or link) and D is a regular diagram for K. Our intention is to decompose D into several simple closed curves. The first step is to draw a small circle with one of the crossing points of D as its centre. This circle intersects D at four points, say, a, b, c, and d, Figure 5.1.1(a). As shown in Figure 5.1.1(b), let us splice this crossing point and connect a and d, and b and c.

Figure 5.1.1

What we have done is to change the original segments ac and bd into the new segments ad and bc. In this way we can remove the crossing point of D that lies within the circle. This operation is called the *splicing* of a knot K (along its orientation) at a crossing point of D. If we perform this splicing operation at every crossing point of D, then we shall remove all the crossing points from D. The end result is that D becomes decomposed into several simple closed curves, Figure 5.1.2(b). These curves are called *Seifert curves*. D, itself, has been transformed into a regular diagram of a link on the plane that possesses no crossing points (i.e., the trivial link). Each of these simple closed curves may now be spanned by a disk.

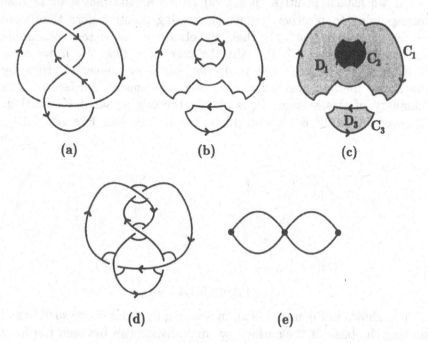

Figure 5.1.2

In the case of Figure 5.1.2(b), by slicing we obtain three disks, D_1, D_2, and D_3, Figure 5.1.2(c). The boundary of D_i is the Seifert curve C_i. In Figure 5.1.2(c), there is a possibility that D_2 may lie on top of D_1, or D_2 may be under D_1. This ambiguity causes some difficulties in Section 3 (see Example 5.3.4). However, these difficulties will be resolved once we prove Theorem 5.4.1. Finally, in order to create a single surface from the various disks, we need to attach to these disks small bands that have been given a single twist. To do this, firstly, take a square acbd and give it a single positive or negative twist, Figure 5.1.3(a) and (b), respectively; these twisted squares are the required bands.

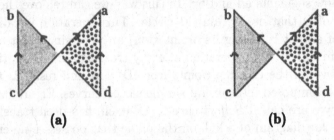

(a)   (b)

Figure 5.1.3

If we attach positive (negative) bands at the places of D that corresponded to positive (negative) crossing points before they were spliced (see Figure 4.5.1), then we obtain a connected, orientable surface F, Figure 5.1.2(d). (In the case of a link, K, if we alter K in such a way that the projection of K is connected, then by the above method we can also obtain a connected surface.) The boundary of this surface, F, is plainly the original knot K. Further, as noted above, F is also an orientable surface (see Exercise 5.1.1). ∎

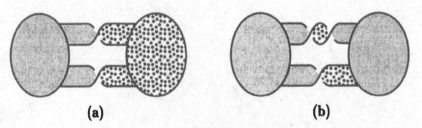

(a)   (b)

Figure 5.1.4

As shown in Figure 5.1.4(a), by shading the front of the surface and dotting the back of the surface, we may distinguish between the front and the back of the surface. This allows us to assign an orientation to the surface. However, as in Figure 5.1.4(b), if one of the bands has a double twist, then it is not possible for us to distinguish between the front and the back.

Exercise 5.1.1. Show that the surface constructed in the proof of Theorem 5.1.1 is an orientable surface.

In general, an orientable, connected surface that has as its boundary an oriented knot (or link) K is called a *Seifert surface* of K. (Due to its origins, maybe, in fact, we should call it a Pontrjagin-Frankl-Seifert surface.) The orientation of F is induced naturally from the orientation of the knot K that forms its boundary. The Seifert surface that was constructed in the above proof depended on the regular diagram, D, of K. Hence, it is more precise to say that is the *Seifert surface formed from* D, a regular diagram of K.

Caveat lector, in the link case, even a seemingly innocuous change of orientation of a component(s) may cause the Seifert surface to change quite substantially.

Suppose, now, that a surface, F, is the Seifert surface of a knot K obtained from the disks and bands as described above. If we shrink (contract) each disk to a point, and at the same time the width of the bands is shrunk, ideally, into quite narrow segments, then from these points and segments a graph in space is formed. Such a graph is called the *Seifert graph* (of a regular diagram D) of K. These graphs, in fact, lie on the plane, i.e., they are plane graphs (cf. Exercise 5.1.2). Figure 5.1.2(e) is the Seifert graph of the figure 8 knot, Figure 5.1.2(a), with vertices (segments) corresponding to the disks (bands, respectively).

Exercise 5.1.2. (a) Show that a Seifert graph is a plane graph. Further, show that it is also a bipartite plane graph. [A graph G (not necessarily plane) is said to be *bipartite* if the set of vertices of G can be divide into two non-empty disjoint subsets, V_1 and V_2, such that every edge of G has one end in V_1 and the other in V_2.]

(b) Show that a graph G is bipartite if and only if every closed path of G consists of an even number of edges, and, in particular, show that a bipartite graph cannot have a loop.

Exercise 5.1.3. Construct Seifert surfaces (obtained from the regular diagrams) for the knots and links in Figure 5.1.5. Also, determine their Seifert graphs. Further, change the orientation of one of the components in Figure 5.1.5(d) and once again determine its Seifert graph. Compare this Seifert graph with the previous one.

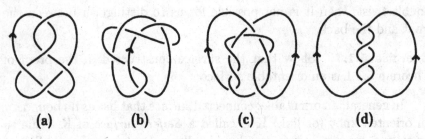

(a)   (b)   (c)   (d)

Figure 5.1.5

§2 The genus of a knot

At this juncture, we ask the reader's indulgence as we now need to consider a well-known, fundamental theorem[11] in topology that is concerned with the classification of surfaces. This theorem states that a closed (i.e., one that is compact and without boundary) orientable surface, F, is topologically equivalent (i.e., homeomorphic) to the sphere with several handles attached to its surface. The number of these handles is called the *genus of* F, and is denoted by g(F).

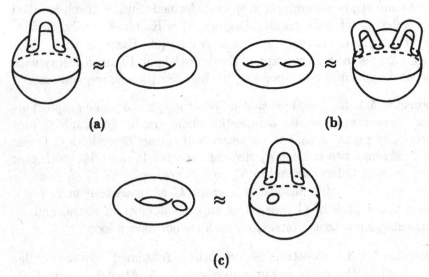

Figure 5.2.1

Example 5.2.1. The surface of genus 1, shown in Figure 5.2.1(a), is called a *torus,* while the surface in Figure 5.2.1(b) has genus 2.

Let us now consider how to calculate the genus of a Seifert surface, F, of a knot. Since F has a boundary, then by the above theorem F is homeomorphic to a sphere with several handles attached, *and further-more,* with a *hole* on the sphere (with handles) for each component of the link, Figure 5.2.1(c). Unfortunately, it is usually not that easy to visualize the Seifert surface of a knot. For example, the Seifert surface of the figure 8 knot shown in Figure 5.1.2(d) is, in fact, topologically the same surface as in Figure 5.2.1(c).

Seifert, by the method we described in the previous section, re-confirmed that such orientable surfaces do exist, and hence in theory, anyway, we may consider the minimum genus of all Seifert surfaces for a given knot K. This minimum genus is called the *genus of* K, denoted by g(K). The genus is a knot invariant (the invariant is defined, as on previous occasions, as the *minimum* one over such genera of a given K). For an arbitrary knot there does exist an algorithm to actually calculate its genus, but it is exceedingly difficult to implement. In truth, to calcu-late the genus of an arbitrary knot is a difficult undertaking. However, for certain types of knots the calculation of the genus is a relatively straightforward matter (cf. Chapter 7 and Chapter 11, Section 5). Al-though the determination of the genus of an arbitrary knot is difficult, to determine the genus of "constructed" orientable surface is quite easy. The calculation relies on a classical invariant, the Euler characteristic.

Theorem 5.2.1.

We may divide a closed orientable surface into α_0 points, α_1 edges, and α_2 faces. Let

$$\chi(F) = \alpha_0 - \alpha_1 + \alpha_2;$$

then $\chi(F)$ is an integer that is independent of how we have divided F; i.e., it is only dependent on F. This integer is called the Euler charac-teristic *of F.*

The Euler characteristic $\chi(F)$ and the genus of F, g(F), are related by means of the following equation:

$$\chi(F) = 2 - 2g(F). \tag{5.2.1}$$

Therefore,

$$g(F) = \frac{2 - \chi(F)}{2}.$$

If F has a boundary, since the boundary is also composed of several points and edges, the above formula becomes

$$\chi(F) = 2 - \mu(F) - 2g(F), \tag{5.2.2}$$

where $\mu(F)$ *is the number of closed curves that make up the boundary of F.*

Example 5.2.2. We can divide the torus with a hole in the manner shown in Figure 5.2.2, so that $\alpha_0 = 7$, $\alpha_1 = 14$, and $\alpha_2 = 6$. It follows from this that $\chi(F) = -1$, and therefore $g(F) = 1$.

Exercise 5.2.1. Show, by suitably dividing it, that the sphere S^2 has Euler characteristic 2.

Let us now apply the above Euler characteristic, (5.2.2), to the Seifert surface that were previously constructed. We may think of the disks and bands of F as a division of F. The points of F in this division are the four vertices of each band. The edges of F are the polygonal curves that constitute the edges of the bands and the boundaries of the disks between the vertex points. The faces of F are the disks and the bands.

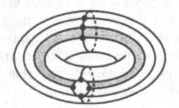

Figure 5.2.2

Exercise 5.2.2. Show that if d is the number of disks and b the number of bands, then $\alpha_0 = 4b$, $\alpha_1 = 6b$, and $\alpha_2 = b + d$.

Therefore, it follows from Exercise 5.2.2 that $\chi(F) = 4b - 6b + b + d = d - b$. Further, $\mu(K)$ is just the number of components of the link K. So from (5.2.2) we obtain that

$$2g(F) = 2 - \mu(K) - \chi(F) = 2 - \mu(K) - d + b,$$

or equivalently,

$$2g(F) + \mu(K) - 1 = 1 - d + b.$$

In the special case when K is a knot, since $\mu(K) = 1$ it follows that

$$2g(F) = 1 - d + b.$$

For the rest of the section let us consider this number, i.e., $1-d+b$. Suppose $\Gamma(D)$ is the Seifert graph constructed from the Seifert surface in Figure 5.1.2(e). Since $\Gamma(D)$ is a plane graph, $\Gamma(D)$ divides S^2 into several domains. (We may think of the sphere S^2 as $\mathbf{R}^2$ with the addition of the point at infinity.) In this partition of S^2, the number of points is d and the number of edges is b. Suppose that the number of faces is f; then from Theorem 5.2.1 and Exercise 5.2.1 we obtain,

$$2 = \chi(S^2) = d - b + f.$$

Therefore,

$$f - 1 = 1 - d + b, \tag{5.2.3}$$

i.e., $1 - d + b$ is equal to the number of faces of this division of S^2, excluding the face that contains the point at infinity, ∞.

Exercise 5.2.3.   Calculate the genus of each Seifert surface of the knots and links in Exercise 5.1.3. Further, with regard to the Seifert graphs obtained from these surfaces, verify that the above formula, (5.2.3), holds.

§3 The Seifert matrix

Suppose that F is a Seifert surface created from the regular diagram, D, of a knot (or link) K, and $\Gamma(D)$ is its Seifert graph. We want to create exactly $2g(F) + \mu(K) - 1$ closed curves[12] that lie on F.

When $\Gamma(D)$ partitions S^2, then we showed in the previous section that $2g(F) + \mu(K) - 1 \ (= 1 - d + b = f - 1)$ is equal to the number of domains (excluding the domain that contains ∞). The boundary of each of these domains (faces) is a closed curve of $\Gamma(D)$. Therefore, we can, from these closed curves, create the closed curves on the Seifert surface.

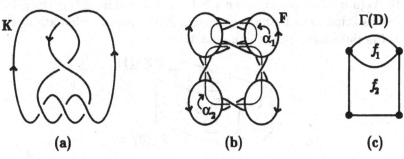

| (a) | (b) | (c) |

Figure 5.3.1

Example 5.3.1. The two closed curves α_1 and α_2, Figure 5.3.1(b) on F, correspond to the boundaries of the two faces, excluding the one that contains ∞, f_1 and f_2 (on S^2), Figure 5.3.1(c), obtained from $\Gamma(D)$.

These $2g(F) + \mu(K) - 1$ $(= m)$ closed curves, it would seem, are nothing but a collection of very ordinary closed curves. However, they will, with a bit of perspicacity, indicate certain characteristics of the knot K that is the boundary of the surface F. [In this case, it is necessary that the Seifert surface is in $\mathbf{R}^3$ (or S^3).] Individually, however, these closed curves are of little interest, but as a collection of closed curves they will provide us with a knot invariant. For example, the knot K in Figure 5.3.1(a) has a Seifert surface of genus 1, from which we can obtain two closed curves, α_1 and α_2. It is quite possible that α_1 and α_2 may have points of intersection; therefore, $\{\alpha_1, \alpha_2\}$ is not a link. But if we lift α_2 slightly above the surface, so that the new curve $\alpha_2^{\#}$ is "parallel" to curve α_2, we can remove these points of intersection, thus making $\{\alpha_1, \alpha_2{}^{\#}\}$ a link.

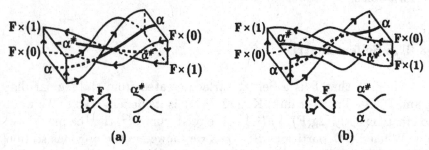

Figure 5.3.2

In order to make this lift precise and slightly easier to understand, let us consider the mathematical construction shown in Figure 5.3.2.

Firstly, we need to thicken F slightly; in other words, create F × $[0,1]$, Figure 5.3.2. Some care needs to be taken during this thickening process, so that both F and the segment $[0,1]$ have the orientations that obey the right-hand rule, Figure 5.3.3.

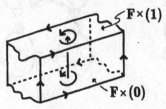

Figure 5.3.3

The original surface F may be thought of as $F \times (0)$, and so we may say that both α_1 and α_2 lie on $F \times (0)$. To be exact α_1 and α_2 should now be called $\alpha_1 \times (0)$ and $\alpha_2 \times (0)$, respectively. For the sake of simplicity we shall retain the original notation, and also for this purpose we shall denote $\alpha_1 \times (1)$ and $\alpha_2 \times (1)$ by $\alpha_1{}^\#$ and $\alpha_2{}^\#$, respectively.

We may assign an orientation to α_1 and α_2 in an arbitrary fashion. These orientations induce, in a natural manner orientations on $\alpha_1^\#$ and $\alpha_2^\#$. This now allows us to calculate the linking number $\mathrm{lk}(\alpha_1, \alpha_2{}^\#)$. It is possible to similarly define the linking numbers $\mathrm{lk}(\alpha_2, \alpha_1{}^\#)$, $\mathrm{lk}(\alpha_1, \alpha_1{}^\#)$, and $\mathrm{lk}(\alpha_2, \alpha_2{}^\#)$. These four linking numbers may be rearranged into the following 2×2 matrix form:

$$M = \begin{bmatrix} \mathrm{lk}(\alpha_1, \alpha_1{}^\#) & \mathrm{lk}(\alpha_1, \alpha_2{}^\#) \\ \mathrm{lk}(\alpha_2, \alpha_1{}^\#) & \mathrm{lk}(\alpha_2, \alpha_2{}^\#) \end{bmatrix}.$$

This matrix M is called the *Seifert matrix* of the knot K in Figure 5.3.1(a). Since the linking numbers themselves are integers, the matrix M is an integer matrix. (It should be noted, however, that the matrix M depends on the orientations of α_1 and α_2; therefore, the matrix is not an invariant of K.)

If the genus of the Seifert surface, F, of a knot (or link) is g(F), then on F there are $2g(F) + \mu(K) - 1 (= m)$ closed curves $\alpha_1, \alpha_2, \ldots, \alpha_m$. Expanding the process outlined above, with arbitrary orientations assigned to these closed curves, we can calculate their various linking numbers. As above, we may formulate them in terms of the entries of a $m \times m$ matrix,

$$M = [\mathrm{lk}(\alpha_i, \alpha_j{}^\#)]_{i,j=1,2,\ldots,m}.$$

Therefore, from the regular diagram, D, of K, we can obtain an integer matrix. However, we should underline that this matrix depends on the orientations of $\alpha_1, \alpha_2, \ldots, \alpha_m$. This matrix is called the *Seifert matrix* of K (*constructed* from a particular regular diagram of K). In general, the linking numbers $\mathrm{lk}(\alpha_i, \alpha_j{}^\#)$ and $\mathrm{lk}(\alpha_j, \alpha_i{}^\#)$ are not equal, so the matrix M is not a symmetric matrix.

Further, in the case when g(F) = 0, the Seifert matrix of K is defined to be the empty matrix (K, as we have already mentioned, is the trivial knot).

Let us now carefully consider several examples to illustrate how to calculate, in practice, the Seifert matrix of a knot.

Example 5.3.2. If we transform the regular diagram of the right-hand trefoil knot to the one in Figure 5.3.4(a), then it is fairly straightforward to see its Seifert surface is the one in Figure 5.3.4(b) and the subsequent Seifert graph is as in Figure 5.3.4(c).

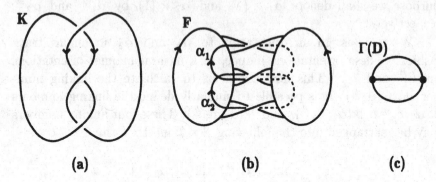

Figure 5.3.4

From Figure 5.3.4(b) it follows that there are two closed curves α_1 and α_2 on the Seifert surface. The mutual relationships between α_1, α_2, $\alpha_1{}^{\#}$ and $\alpha_2{}^{\#}$ are shown in Figure 5.3.5(a) $\sim$ (d).

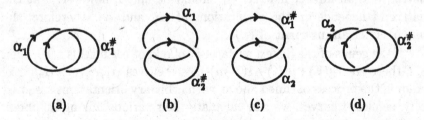

Figure 5.3.5

From these four diagrams it follows that

$$\mathrm{lk}(\alpha_1, \alpha_1{}^{\#}) = -1, \ \mathrm{lk}(\alpha_2, \alpha_1{}^{\#}) = 1, \ \mathrm{lk}(\alpha_2, \alpha_2{}^{\#}) = -1,$$

with the linking number of the other case equal to 0.

Therefore, the Seifert matrix for the right-hand trefoil knot is

$$M = \begin{bmatrix} -1 & 0 \\ 1 & -1 \end{bmatrix}.$$

Example 5.3.3. If in a similar manner we consider the Seifert matrix of the left-hand trefoil knot, we obtain the following matrix:

$$M = \begin{bmatrix} 1 & -1 \\ 0 & 1 \end{bmatrix}.$$

Example 5.3.4.   Let us now consider the knot, K, in Figure 5.3.6(a).

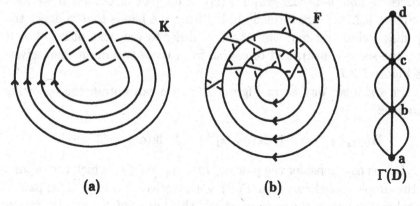

(a)   **(b)**

Figure 5.3.6

Various Seifert surfaces can be constructed from this diagram. In particular, we shall consider the two cases in Figure 5.3.7(a) and (b).

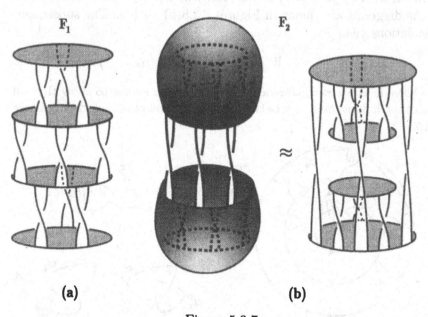

(a)   **(b)**

Figure 5.3.7

Since the Seifert surface of K has genus 3, then its Seifert matrix is a 6×6 matrix, and on the surface correspondingly there are 6 closed curves. First, let us consider case (a). To try to avoid too much confusion, we shall build up the linking number formulae pair upon pair.

Let us begin with the pair α_1 and α_2 on F_1, which correspond to the vertices a and b of the graph $\Gamma(D)$. This pair of curves, if we look at Figure 5.3.7(a), are formed on F_1 from the bands that connect the top disk (which we may call the first disk) to the next disk below it (i.e., the second disk), and of course from these two disks themselves; see Figure 5.3.7.

In a similar way as in Figure 5.3.5, we can obtain the following formulae:

$$\text{lk}(\alpha_1, \alpha_1{}^{\#}) = -1, \; \text{lk}(\alpha_2, \alpha_1{}^{\#}) = 1, \; \text{lk}(\alpha_2, \alpha_2{}^{\#}) = -1.$$

Let us now consider the pair α_3 and α_4, on F_1, which correspond to the simple closed curves in $\Gamma(D)$ with vertices b and c. This pair of closed curves lies on the the second and third disks of F_1 and the bands that connect these two disks. We leave as an exercise the calculation of the linking numbers for this pair, namely between α_3, $\alpha_3^{\#}$, α_4, and $\alpha_4^{\#}$. The next step is for us to calculate the mutual linking numbers between the two pairs of closed curves, i.e., α_1, α_2 and α_3, α_4, some of the diagrams are shown in Figures 5.3.8(b) $\sim$ (d). The subsequent calculations yield

$$\text{lk}(\alpha_3, \alpha_1{}^{\#}) = 1, \; \text{lk}(\alpha_4, \alpha_1{}^{\#}) = -1, \; \text{lk}(\alpha_4, \alpha_2{}^{\#}) = 1.$$

We leave it as a straightforward exercise for the reader to show that all the other linking numbers between these two pairs of closed curves are zero.

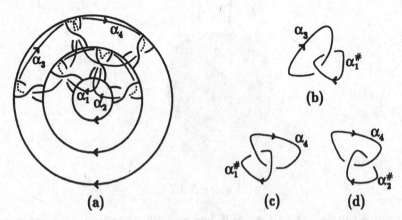

Figure 5.3.8

If we continue in this vein, we next have to add the final pair of closed curves and calculate the relevant linking numbers. This is reasonably straightforward and so we shall, without direct computation, give the subsequent matrix, the Seifert matrix of K, leaving it as an exercise for the reader to check that the linking numbers are as printed.

$$M = \begin{bmatrix} -1 & 0 & 0 & 0 & 0 & 0 \\ 1 & -1 & 0 & 0 & 0 & 0 \\ 1 & 0 & -1 & 0 & 0 & 0 \\ -1 & 1 & 1 & -1 & 0 & 0 \\ 0 & 0 & 1 & 0 & -1 & 0 \\ 0 & 0 & -1 & 1 & 1 & -1 \end{bmatrix}.$$

For the second case, Figure 5.3.7(b), similar calculations lead to the following Seifert matrix M' from the Seifert surface F_2 :

$$M' = \begin{bmatrix} -1 & 0 & -1 & 1 & 0 & 0 \\ 1 & -1 & 0 & -1 & 0 & 0 \\ 0 & 0 & -1 & 0 & 0 & 0 \\ 0 & 0 & 1 & -1 & 0 & 0 \\ 0 & 0 & 1 & 0 & -1 & 0 \\ 0 & 0 & -1 & 1 & 1 & -1 \end{bmatrix}.$$

Exercise 5.3.1. Determine the Seifert matrix obtained from the regular diagram D of the knot in Example 5.3.1, with an arbitrary (i.e., at the reader's discretion) orientation assigned to α_1 and α_2.

Exercise 5.3.2. Determine the Seifert matrix for the knots in Figure 5.1.5(a) and (b). In particular, for the knot in Figure 5.1.5(b), find two Seifert matrices by applying the method explained in Example 5.3.4.

As was noted in the above example, the Seifert matrix of a knot is not unique. In fact, since we have fixed neither the orientation nor the order of the closed curves $\alpha_1, \alpha_2, \ldots, \alpha_m$, even by the seemingly minor adjustment of changing the order, we can cause the Seifert matrix to change. Therefore, in order to obtain an invariant of a knot from a Seifert matrix, we need to examine the relationship between the Seifert matrices of the same knot. The concept that is being alluded to in the preceding sentence is the S-equivalence of two square matrices.

§4 S-equivalence of Seifert matrices

The construction of the Seifert matrix outlined above depends on the regular diagram we use. Hopefully the following statement is now second nature to the reader: We may transform one regular diagram into another equivalent regular diagram by applying the Reidemeister moves several times. Therefore, our next course of action, if we wish to use the Seifert matrices to define knot invariants, is to carefully examine the effect of the Reidemeister moves on the Seifert matrix.

Theorem 5.4.1.

Two Seifert matrices, obtained from two equivalent knots (or links), can be changed from one to the other by applying, a finite number of times, the following two operations, Λ_1 and Λ_2, and their inverses:

$$\Lambda_1 : M_1 \to PM_1P^T,$$

where P is an invertible integer matrix, with $\det P = \pm 1$ ($\det P$ is just the usual determinant of P), and P^T denotes the transpose matrix of P.

$$\Lambda_2 : M_1 \to M_2 = \begin{bmatrix} & & * & 0 \\ & M_1 & \vdots & \vdots \\ & & * & 0 \\ 0 & \cdots & 0 & 0 & 1 \\ 0 & \cdots & 0 & 0 & 0 \end{bmatrix} \text{ or } \begin{bmatrix} & & 0 & 0 \\ & M_1 & \vdots & \vdots \\ & & 0 & 0 \\ * & \cdots & * & 0 & 0 \\ 0 & \cdots & 0 & 1 & 0 \end{bmatrix},$$

where $*$ denotes an arbitrary integer.

The above mathematical argot is essential for the theorem to be precise, but let us peel away some of this terminology and try to understand exactly what effect the two operations will have on a matrix.

The operation Λ_1 either interchanges two rows, say i^{th} and j^{th} rows, and then interchanges the ith and j^{th} columns; or it adds k times the ith row to the jth row, and then adds k times the ith column to the jth column. We shall call this operation an *elementary symmetric matrix operation.* The operation has been defined in such a way that it corresponds to the change of order or the change of orientation of the closed curves mentioned above and others.

The operation Λ_2, on the other hand, is a matrix operation that is particular to knot theory. This operation has been defined so that it corresponds to the change in the genus of the Seifert surface due to a Reidemeister move, i.e., it makes the Seifert matrix either smaller or larger.

Exercise 5.4.1. In Example 5.3.2 reverse the orientation of α_1 (and hence, $\alpha_1^\#$) and then determine the new Seifert matrix M'. Compare

M and M', and confirm that M' is obtained from M by multiplying the first row and first column by (-1).

Definition 5.4.1. Two square matrices M, M' obtained one from the other by applying the operations Λ_1, Λ_2 and the inverse Λ_2^{-1} a finite number of times, are said to be *S-equivalent,* and are denoted by M $\overset{S}{\sim}$ M'. (The "S," of course, stands for Seifert.)

For convenience's sake, we say two matrices are Λ_1-*equivalent* if one is obtained from the other by applying the operation Λ_1 a finite number of times.

Now, two Seifert matrices obtained from two equivalent knots (or links) are, by Theorem 5.4.1, S-equivalent. Before we proceed with the proof of Theorem 5.4.1, to avoid a situation as described in the next paragraph from occurring, we would like to generalize and refine some of our previous concepts.

A Seifert surface is, by definition, a *connected surface.* If a given regular diagram D is not connected, we need to transform it into a connected regular diagram, $\hat{D}$, by applying the Reidemeister move Ω_2. Hence, a suitable connected surface can now be constructed. However, it is possible that as we apply subsequent Reidemeister moves to transform our original regular diagram to an equivalent regular diagram, we shall encounter, within the intermediary regular diagrams, one that is not connected, thus returning to our original problem. So to stop chasing our tails, it is better to *redefine* the Seifert matrix, making it independent of whether the Seifert surface is connected or disconnected.

Therefore, let D be a regular diagram with p connected components D(1), D(2), ... D(p), ($p \geq 1$). We can, using the methods already described, construct Seifert surface F(i) for each D(i) and subsequently a Seifert matrix M(i) from F(i), $i = 1, 2, \ldots, p$.

Definition 5.4.2. The Seifert matrix M of a disconnected (Seifert) surface F(1) $\cup$ F(2) $\cup \ldots \cup$ F(p) is defined to be the direct sum of M(1), M(2), ... M(p) and the zero matrix O_{p-1} of order $p - 1$, i.e.,

$$
M = \begin{bmatrix}
M(1) & & & \\
& M(2) & & O \\
& & \ddots & \\
O & & M(p) & \\
& & & O_{p-1}
\end{bmatrix}.
$$

Next, we shall show the following proposition is a straightforward consequence of this definition.

Proposition 5.4.2.

Let $\widehat{F}$ be the connected surface obtained from $F(1) \cup F(2) \cup \ldots \cup F(p)$ by adding two bands with an opposite twist, see Figure 5.4.1, between $F(i)$ and $F(i+1)$, $i = 1, 2, \ldots, p-1$. (So, $\widehat{F}$ is a Seifert surface constructed from a connected diagram $\widehat{D}$.) Then the matrix $\widehat{M}$ obtained from $\widehat{F}$ is Λ_1-equivalent to the matrix defined in Definition 5.4.2; hence, $\widehat{M} \overset{S}{\sim} M$.

F(1)   **F(2)**   **F(p)**

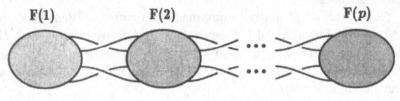

Figure 5.4.1

Proof

On a pair of these new bands, place a new simple closed curve α_i, $i = 1, 2, \ldots, p-1$, see Figure 5.4.2.

F(i)   **F(i+1)**

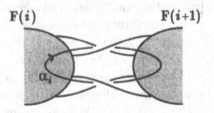

Figure 5.4.2

Since for each original simple closed curve $\alpha_{j,k}$ constructed on $F(j)$, $\mathrm{lk}(\alpha_{j,k}, \alpha_i^{\#}) = 0$ and $\mathrm{lk}(\alpha_i, \alpha_{j,k}^{\#}) = 0$ [of course $\mathrm{lk}(\alpha_i, \alpha_i^{\#}) = 0$], we have that $\widehat{M}$ is Λ_1-equivalent to M. (*Nota bene*, the only difference between $\widehat{M}$ and M is either a change of the numbering of the $\alpha_{j,k}$ or a change in the orientation.) ∎

We are now in a better position to proceed with the proof of Theorem 5.4.1.

Proof of Theorem 5.4.1.

To prove the theorem, we need first to look at how the Seifert surface changes when we apply each of the Reidemeister moves, and secondly, as a consequence, to examine how the Seifert matrix changes. In fact, due to the next proposition, we may *restrict* the proof to consider only local changes of the surface.

Proposition 5.4.3.

Let D be a regular diagram and let D′ be the regular diagram obtained from D by applying only a single Reidemeister move on it. Further, let F be a Seifert surface constructed from D. Similarly, we can construct a F′ from D′, but this can be done so that F and F′ differ only at the parts that are affected by the Reidemeister move. (In other words, F and F′ are identical except at a few places.) Then the Seifert matrices M and M′ obtained from F and F′, respectively, are S-equivalent. (In fact, by the very construction of F′, M′ is identical to M, except at a few rows and columns.)

The proof of Theorem 5.4.1 will be complete if, in addition to the above proposition, we can prove that two different Seifert surfaces constructed from the same regular diagram have S-equivalent Seifert matrices. This actually falls out from the next proposition.

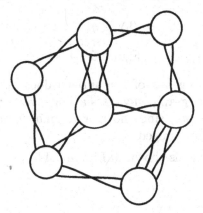

Figure 5.4.3

Proposition 5.4.4.

Let F be one of the Seifert surfaces constructed from a regular diagram D, and let M_F the subsequent Seifert matrix of F. Then there exists a diagram D_0 such that

  (1) D_0 is equivalent to D.
  (2) Only one Seifert surface, F_0, is constructed from D_0; i.e., all disks are on the <u>same</u> level, see Figure 5.4.3. (Such a surface F_0 is sometimes called flat.)

(3) *The Seifert matrix of* F_0, M_{F_0}, *is* Λ_1-*equivalent to* M_F, *and hence,* $M_{F_0} \overset{S}{\sim} M_F$.

Exercise 5.4.2. Show that Proposition 5.4.3 and 5.4.4 imply Theorem 5.4.1.

<u>Proof of Proposition 5.4.3.</u>

The idea of the proof is quite straightforward, it requires only checking several possible cases. So we shall look at some of these cases and leave the rest as exercises for the reader.

The single Reidemeister move applied is Ω_1.

With regard to this Reidemeister move, we shall consider two cases:

(i) We increase the number of bands and disks by only one of each, as in Figure 5.4.4(b).

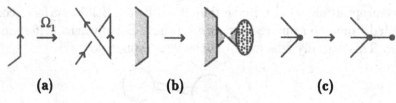

    (a)   **(b)**   **(c)**

Figure 5.4.4

In this case the genus of the surface does not change, and the corresponding Seifert graph only adds a single vertex and an edge, Figure 5.4.4(c). Therefore, since no new domain is created, the Seifert matrix will remain unchanged.

(ii) The move is as shown in Figure 5.4.5.

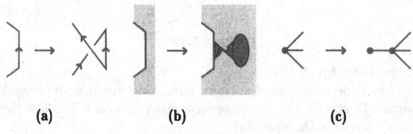

    (a)   **(b)**   **(c)**

Figure 5.4.5

However, this can be dealt with along similar lines to (i), and so the Seifert matrix remains unchanged.

The single Reidemeister move applied is Ω_2.

Since the possibilities depend on how we assign the orientation to each segment, there are several cases to consider. We shall only look at some typical cases.

(i) In the case of Figure 5.4.6 the number of disks does not change, but the number of bands increases by 2.

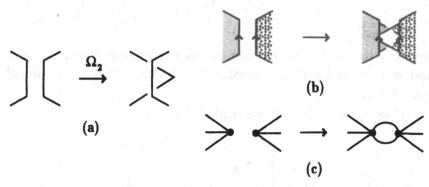

Figure 5.4.6

If the original Seifert surface F is not connected, but the new surface F′ is connected, then it follows from Proposition 5.4.2 that $M \overset{S}{\sim} M'$. So suppose that F is connected. Then the genus of F′ is given by $g(F') = g(F) + 1$. Therefore, F′ compared with F has an extra 2 closed curves, which we will denote by α' and α'', Figure 5.4.7.

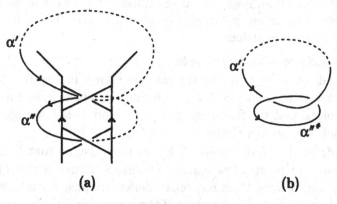

Figure 5.4.7

We may suppose that the newly created two bands are the final bands that connect the disks D′ and D″. (If necessary, we may change the numbering of the closed curves.) The new matrix, M′, in compari-

son with M, has an extra two rows and columns, as shown below. Note
that $\mathrm{lk}(\alpha', \alpha''^{\#}) = 1$, see Figure 5.4.7(b).

$$
M' = \begin{bmatrix}
 & & & b_1 & 0 \\
 & M & & \vdots & \vdots \\
 & & & b_m & 0 \\
b'_1 & \cdots & b'_m & b & 1 \\
0 & \cdots & 0 & 0 & 0
\end{bmatrix},
$$

where b_i is the linking number between the closed curve α_i on F
and $\alpha'^{\#}$, b'_i is the linking number between α' and $\alpha_i^{\#}$, and finally
$b = \mathrm{lk}(\alpha', \alpha'^{\#})$.

If we apply Λ_1 to M' we shall obtain,

$$
M'' = \begin{bmatrix}
 & & & b_1 & 0 \\
 & M & & \vdots & \vdots \\
 & & & b_m & 0 \\
0 & \cdots & 0 & 0 & 1 \\
0 & \cdots & 0 & 0 & 0
\end{bmatrix}.
$$

What we have shown is that the Seifert matrix M', constructed
from F′, can be obtained by first performing Λ_2 to M, the Seifert matrix of F, and then performing Λ_1 several times. Therefore, these Seifert
matrices obtained from F and F′ are S-equivalent. The other variations
on this Reidemeister move can also be shown to give S-equivalence of
the relevant Seifert matrices.

The single Reidemeister move applied is Ω_3.

We shall consider only the typical case shown in Figure 5.4.8(a).
In Figure 5.4.8, the numbers $1, 2, \ldots, 6$ and the letters a, b, c indicate
crossing points, and the Seifert graphs $\Gamma(D)$ and $\Gamma(D')$ are identical,
except inside the broken circles.

Now, from the construction of F′ we may assume that F and F′
are identical, except at a few places. The exact nature of these places
is best visualized from their respective Seifert graphs, since then they
correspond precisely to the interiors of the broken circles, similar to the
type shown in Figure 5.4.8(c). It is then easy, from Figure 5.4.8(c), to
see that F and F′ have the same number of bands, but F has 2 more
disks than F′. [Note that a disk corresponds to a vertex and a band
corresponds to an edge in $\Gamma(D)$ or $\Gamma(D')$.]

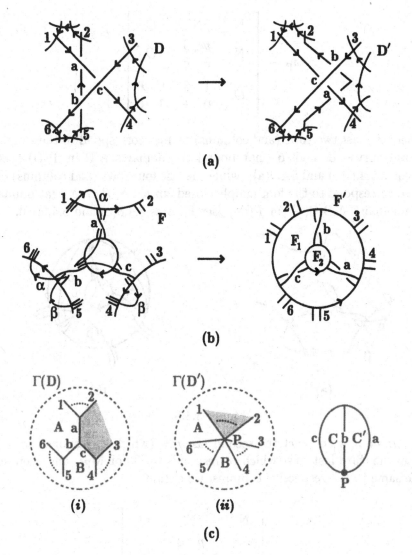

Figure 5.4.8

If we regard the shaded area as corresponding to the unbounded region, then when a disk F_2 lies over F_1, see Figure 5.4.9, the Seifert matrices may be seen to be of the forms,

$$M_F = \begin{bmatrix} N & N' \\ N'' & \begin{matrix} p & 0 \\ -1 & q \end{matrix} \end{bmatrix}$$

and

$$
M_{F'} = \begin{bmatrix} N & N' & O \\ N'' & \begin{matrix} p & 0 \\ 0 & q \end{matrix} & O \\ O & \begin{matrix} 1 & 0 \\ 0 & 1 \end{matrix} & \begin{matrix} 0 & 0 \\ 1 & 0 \end{matrix} \end{bmatrix},
$$

where the last two rows (and columns) of M_F correspond to the simple closed curves α and β that bound the domains A,B in $\Gamma(D)$ [see Figures 5.4.8 (c) and 5.4.9(a)], while the last four rows (and columns) of $M_{F'}$ correspond to the four simple closed curves α, β, γ, δ that bound the domains A,B,C, C' in $\Gamma(D')$ [see Figures 5.4.8(c) and 5.4.9(b)].

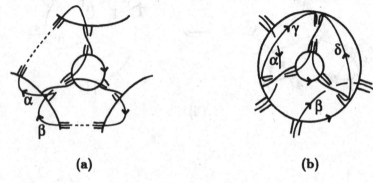

(a)   (b)

Figure 5.4.9

If we now subtract the second last row (which corresponds to γ) from the third last row (which corresponds to β) of $M_{F'}$, and then do the same to the respective columns, we obtain

$$
M_{F'} = \begin{bmatrix} N & N' & O \\ N'' & \begin{matrix} p & 0 \\ -1 & q \end{matrix} & O \\ O & \begin{matrix} 1 & 0 \\ 0 & 0 \end{matrix} & \begin{matrix} 0 & 0 \\ 1 & 0 \end{matrix} \end{bmatrix}.
$$

This matrix, clearly, may be reduced to M_F by applying Λ_2^{-1}, and therefore, $M_F \overset{S}{\sim} M_{F'}$.

We leave it as an exercise for the reader to show the same argument works when F_2 lies under F_1.

Exercise 5.4.3.   Prove the remaining case, see Figure 5.4.10, required in the proof of Proposition 5.4.3.

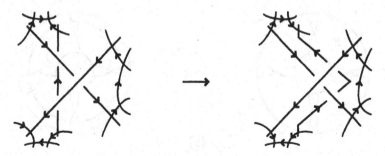

Figure 5.4.10

Proof of Proposition 5.4.4.

We do not wish to burden the reader with yet another turgid proof, and so we only sketch a proof by means of diagrams, leaving the reader to flesh out the details at leisure.

Let us work with the surface shown in Figure 5.4.11.

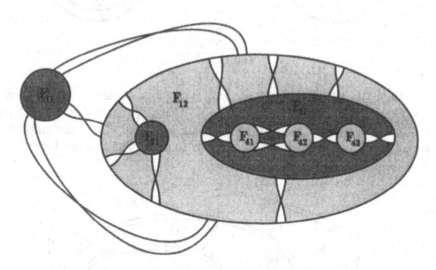

Figure 5.4.11

The ambiguity, and hence the problem arises, when "within" a disk there are other disks. In our example, Figure 5.4.11, there are two such disks, F_{12} and F_{31}. The strategy involved in order to prove this proposition is to replace such a disk by a narrow disk, obtained by applying Reidemeister moves. This will be done in such a way that if F_2 lies over F_1; then the resultant disk lies under all the bands connecting

F_1 and F_2, Figure 5.4.12(a). However, if F_2 lies under F_1 then the narrow disk lies over the bands, Figure 5.4.12(b).

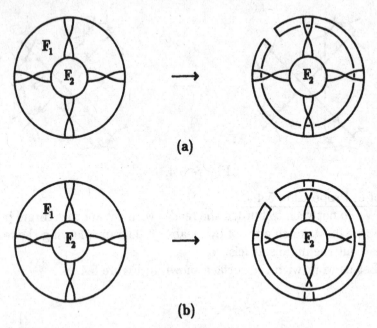

(a)

(b)

Figure 5.4.12

Repeating this process, we will eventually obtain a flat surface. For example, in Figure 5.4.11 suppose that F_{21} lies over F_{12}, F_{31} lies under F_{12}, and F_{41}, F_{42}, F_{43} all lie over F_{31}. Then the suitably transformed (we leave it as an exercise to show this can be done) flat surface, F_0, is shown in Figure 5.4.13.

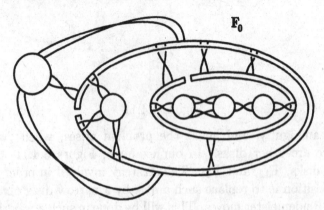

Figure 5.4.13

It is easy to see that F_0 has the same Seifert matrix as F (up to Λ_1-equivalence), see Figure 5.4.14.

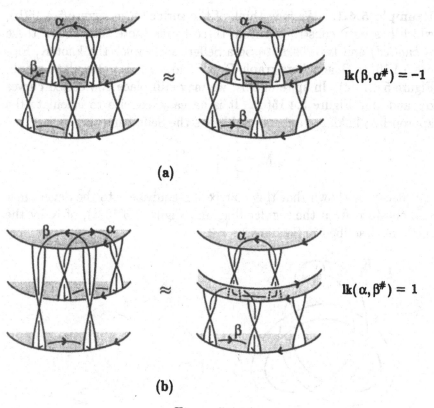

$$\mathrm{lk}(\beta, \alpha^{\#}) = -1$$

(a)

$$\mathrm{lk}(\alpha, \beta^{\#}) = 1$$

(b)

Figure 5.4.14

Exercise 5.4.4. Show that the Seifert matrices M, M' in Example 5.3.4 are S-equivalent. (In fact, they are Λ_1-equivalent.)

From a regular diagram of a knot (or link) K we can create a Seifert surface, and then from this a Seifert matrix. However, to calculate the Seifert matrix it may be possible to avoid this process, for on occasion we may construct a Seifert surface that consists of disks and bands, and on which we place m $(= 2g(F) + \mu(K) - 1)$ closed curves. However, the Seifert graph Γ obtained from F may *not* be a plane graph. Therefore, it is possible that what has gone before does not hold in this case. However, we bring to our aid the next theorem.

Theorem 5.4.5 [Tr].

Suppose that M_1 and M_2 are two Seifert matrices obtained from Seifert surfaces F_1 and F_2 of K. Then M_1 and M_2 are S-equivalent.

We now give an example of the type of Seifert surface alluded to in above discussion.

Example 5.4.1. We may think of the surface F in Figure 5.4.15(b), which has been constructed from three bands (note that one of these is *knotted*) and two disks, to be a Seifert surface for the knot K, Figure 5.4.15(a). The Seifert graph Γ of F, however, is not a plane graph, Figure 5.4.15(c). In spite of this, we may still place two closed curves α_1 and α_2, Figure 5.4.15(d). It is an easy exercise to calculate the appropriate linking numbers, and hence the Seifert matrix,

$$M = \begin{bmatrix} -1 & 0 \\ 1 & -1 \end{bmatrix}.$$

It can be shown that this matrix is S-equivalent to the Seifert matrix obtained from the regular diagram, Figure 5.4.15(a), of K by the methods described in Sections 1 $\sim$ 3.

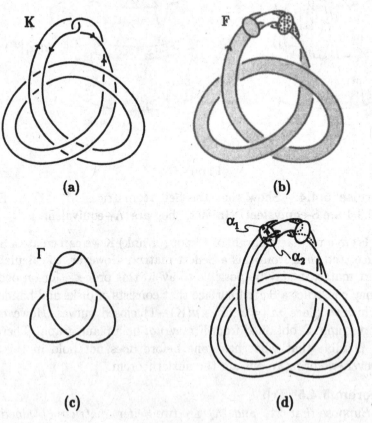

(a)

(b)

(c)

(d)

Figure 5.4.15

So, to be exact, the Seifert matrix should be said to be constructed from the Seifert surface, F, of a knot K. In general, however, we shall simply say that is the Seifert matrix of K. In this vein, we may rewrite Theorem 5.4.5 as folows: Two Seifert matrices of K are S-equivalent. However, we should make it quite clear that when we say a Seifert matrix obtained from a regular diagram D of K, we have in mind that this Seifert matrix has been constructed from D using the methods outlined in Sections 1 $\sim$ 3.

We shall round of this section by proving two properties of Seifert matrices. We shall denote by M_K the Seifert matrix of a knot (or link) K.

Proposition 5.4.6.

Suppose that K is an oriented knot (or link) and $-K$ *is the knot with the reverse orientation to K (in the case, of a link, the orientation is reversed on all the components). Then* $M_{-K}\overset{S}{\sim}M_K^T$, *where* M_K^T *is the transpose matrix of* M_K.

Proof

If we suppose that D is a regular diagram of K, we may take as a regular diagram D' for $-K$, the regular diagram D with all the orientations reversed. Therefore, the orientations of the subsequent Seifert surfaces are completely opposite. Hence, the under and over relations for α_i and $\alpha_j^{\#}$ are also completely reversed. The Seifert matrix obtained from D' is therefore just the transpose of that from D. It follows from Theorem 5.4.1 (or Theorem 5.4.5) that $M_{-K}\overset{S}{\sim}M_K^T$.

∎

Proposition 5.4.7.

Suppose that K^* *is the mirror image of a knot (or link) K, then* $M_{K^*}\overset{S}{\sim}-M_K^T$.

Proof

We can obtain a regular diagram D^* of K^* from K by changing the under- and over-crossing segments at each of the crossing points. Therefore, since the under and over relations for the closed curves that follow from D and D^* are completely reversed, $M_{K^*}\overset{S}{\sim}-M_K^T$. (Compare this with Examples 5.3.2 and 5.3.3.)

∎

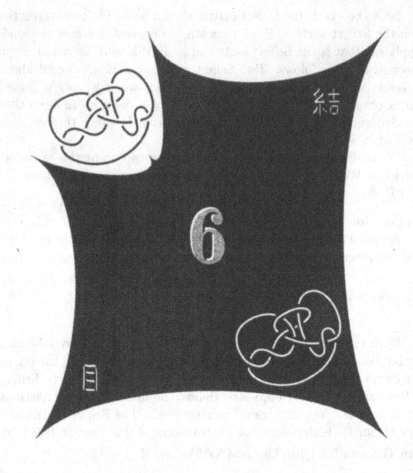

Invariants from the Seifert Matrix

In order to find a knot (or link) invariant from a Seifert matrix, we need to look for something that will not change under the operations Λ_1 and $\Lambda_2^{\pm 1}$, defined in Theorem 5.4.1. We will see in this chapter that the Alexander polynomial is such an invariant. The Alexander polynomial is not the only important invariant that we can extricate from the Seifert matrix, the signature of a link can also be defined from it. In addition to defining these two invariants we shall, in this chapter, prove some of their basic characteristics. *Nota bene*, throughout this chapter we shall assume all the knots and links are *oriented*.

§1 The Alexander polynomial

For a mathematician it is natural to ask, since we have such a nice tool as the Seifert matrix, what matrix properties do we know that, via Λ_1 and Λ_2, might yield a knot (or link) invariant.

Exercise 6.1.1. Find an example that shows that the determinant, det M, of the Seifert matrix M of a knot K is *not* a knot invariant.

However, we should not discard the idea of using the determinant. Let us, first, symmetrize the matrix M to form the matrix sum $M + M^T$. If we now look at the absolute value of the determinant of $M + M^T$, this does lead to a link invariant.

Proposition 6.1.1.

If M is the Seifert matrix of knot (or link) K, then $\left| \det(M + M^T) \right|$ *is an invariant of the knot K. This invariant is called the* determinant *of K.*

Exercise 6.1.2. Find a proof of Proposition 6.1.1. (Hint: Show that the determinant does not change its value if we apply the operations Λ_1 and $\Lambda_2^{\pm 1}$.)

We will prove later, in Chapter 11, Section 2, that the determinant of a knot (or link) K is completely independent of the orientation assigned to K.

This invariant, the determinant of a knot, is quite an old invariant. One of its useful properties is that since the determinant of the trivial knot is 1 (we define the determinant of the empty matrix to be 1), it can and has, over the years, been used to prove that certain knots are not the trivial knot.

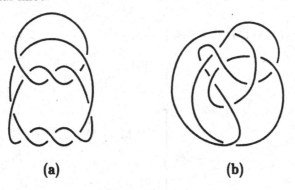

(a)   (b)

Figure 6.1.1

Example 6.1.1.

Since the determinant of the trefoil knot is 3, it is not equivalent to the trivial knot. There are knots, however, that have determinant 1 but are not equivalent to the trivial knot; Figures 6.1.1(a) and (b) are such examples.

The proof of the following proposition we shall postpone until later, since the proof is an easy consequence of Proposition 6.3.1, which is proven a bit later.

Proposition 6.1.2.

Suppose that M is the Seifert matrix of a knot (but not a link) K, then

$$\det(M - M^T) = 1.$$

At this stage, we would like to ask the reader's indulgence as we jump directly from the above determinant to consider the polynomial,

$$\det(M - tM^T),$$

which resembles the characteristic polynomial of M. The determinant is now a polynomial with indeterminate t. The next logical step is to examine how this polynomial changes when we apply Λ_1 and $\Lambda_2^{\pm 1}$.

Firstly, since $\det P = \det P^T = \pm 1$,

$$\det(\Lambda_1(M - tM^T)) = \det[P(M - tM^T)P^T]$$
$$= \det(M - tM^T). \tag{6.1.1}$$

Therefore, it is not affected by the operation Λ_1. However, if we apply Λ_2,

$$\det(\Lambda_2(M - tM^T)) = \det \begin{bmatrix} & & & b_1 & 0 \\ & M - tM^T & & \vdots & \vdots \\ & & & b_m & 0 \\ -b_1 t & \cdots & -b_m t & 0 & 1 \\ 0 & \cdots & 0 & -t & 0 \end{bmatrix}$$

$$= \det \begin{bmatrix} & & & b_1 & 0 \\ & M - tM^T & & \vdots & \vdots \\ & & & b_m & 0 \\ 0 & \cdots & 0 & 0 & 1 \\ 0 & \cdots & 0 & -t & 0 \end{bmatrix}$$

$$= t \det(M - tM^T). \tag{6.1.2}$$

Similarly, we can obtain $\det(\Lambda_2^{-1}(M_2 - tM_2^T)) = t^{-1} \det(M_1 - tM_1^T)$.

These three formulae lead us to the following theorem.

Theorem 6.1.3.

Suppose that M_1 *and* M_2 *are the Seifert matrices for a knot (or link)* K. *Further, if* r *and* s *are, respectively, the orders of* M_1 *and* M_2, *then the following equality holds:*

$$t^{-\frac{r}{2}} \det(M_1 - tM_1^T) = t^{-\frac{s}{2}} \det(M_2 - tM_2^T).$$

Therefore, if M is a Seifert matrix of K and its order is k, then

$$t^{-\frac{k}{2}} \det(M - tM^T)$$

is an invariant of K. This invariant is known as the *Alexander polynomial* of K and is denoted by $\Delta_K(t)$.

It follows directly from our previous discussions that $k = 2g(F) + \mu(K) - 1$, where as before F is the Seifert surface from which we have constructed M, and $\mu(K)$ is the number of components of the link K. In most cases, $\Delta_K(t)$ has some terms with a negative exponent; however, if we multiply $\Delta_K(t)$ by a suitable factor, then we can obtain a polynomial with only positive exponents. Sometimes it is preferable to work with such an interpretation of $\Delta_K(t)$. If K is a link with an even number of components, then k is odd. Therefore, for such links $\Delta_K(t)$ is a polynomial with terms as powers of $t^{\frac{1}{2}} (= \sqrt{t})$ or $t^{-\frac{1}{2}} (= \frac{1}{\sqrt{t}})$. In these cases we define $(t^{\frac{1}{2}})^2 = t$. [In Appendix (II) we tabulate the Alexander polynomial of all prime knots with up to 8 crossings.]

We shall next prove an important property of the Alexander polynomial.

Theorem 6.1.4.

Suppose K is a knot; then $\Delta_K(t)$ *is a symmetric Laurent polynomial, i.e.,*

$$\Delta_K(t) = a_{-n}t^{-n} + a_{-(n-1)}t^{-(n-1)} + \ldots + a_{n-1}t^{n-1} + a_n t^n$$

and

$$a_{-n} = a_n, \; a_{-(n-1)} = a_{n-1}, \ldots, a_{-1} = a_1. \tag{6.1.3}$$

[The more general link case is considered in Exercise 6.2.4(2).]

Proof

Suppose that M is a Seifert matrix of K and k is the order of M. Since K is a knot, k is necessarily even. Therefore,

$$\Delta_K(t^{-1}) = t^{\frac{k}{2}} \det(M - t^{-1}M^T) = t^{-\frac{k}{2}} \det(tM - M^T)$$
$$= (-1)^k t^{-\frac{k}{2}} \det(M^T - tM) = t^{-\frac{k}{2}} \det(M - tM^T)^T$$
$$= t^{-\frac{k}{2}} \det(M - tM^T) = \Delta_K(t).$$

It is now easy to see that (6.1.3) follows directly from this.

∎

Proposition 6.1.5.

$|\Delta_K(-1)|$ *is equal to the determinant of a knot K.*

Proof

$$|\Delta_K(-1)| = \left| (-1)^{-\frac{k}{2}} \det(M + M^T) \right|$$
$$= \left| \det(M + M^T) \right|.$$

∎

Example 6.1.2. If K is a trivial knot, then $\Delta_K(t) = 1$.

Example 6.1.3. If K is the right-hand trefoil knot (cf. Example 5.3.2), then

$$\Delta_K(t) = t^{-1}(M - tM^T) = t^{-1} \det \begin{bmatrix} -(1-t) & -t \\ 1 & -(1-t) \end{bmatrix}$$
$$= t^{-1} - 1 + t.$$

Exercise 6.1.3. Evaluate the Alexander polynomial of the knots in Exercise 5.3.2 and Example 5.3.4.

§2 The Alexander-Conway polynomial

The reader will soon find, by experimenting with the above procedure, that if we wish to use the Alexander polynomial to obtain at least a partial knot table, the above procedure is quite cumbersome. However, due to the constant state of flux in knot theory and its interaction with other disciplines, the above problem can be obviated.

In the late 1950s and the 1960s, computers were transformed from a research project into a research tool. Although the number-crunching abilities of computers were of tremendous advantage, an extra impetus was still required to make the Alexander polynomial more computer friendly. This spark of ingenuity was provided by J.H. Conway in the late 1960s, when he devised an extremely efficient mechanical procedure to compute the Alexander polynomial. (With hindsight, if we carefully reread Alexander's original paper, it is possible to glean from it Conway's method. So perhaps, rather like in the case of fractals, this is a case of technology catching up with mathematical theory.)

Definition 6.2.1. Given an oriented knot (or link) K, then we may assign to it a Laurent polynomial, $\nabla_K(z)$, with a fixed indeterminate z, by means of the following two axioms:

Axiom 1 If K is the trivial knot, then we assign $\nabla_K(z) = 1$.

Axiom 2 Suppose that D_+, D_-, D_0 are the regular diagrams, respectively, of the three knots (or links), K_+, K_-, K_0. These regular diagrams are exactly the same *except* at a neighbourhood of one crossing point. In this neighbourhood, the regular diagrams differ in the manner shown in Figure 6.2.1. (Note: In the case of D_+ (D_-) within this neighbourhood, there exists only a positive (negative) crossing.)

$$\mathbf{D_+} \qquad\qquad \mathbf{D_-} \qquad\qquad \mathbf{D_0}$$

Figure 6.2.1

Then the Laurent polynomials of the three knots (or links) are related as follows:

$$\nabla_{K_+}(z) - \nabla_{K_-}(z) = z\nabla_{K_0}(z). \qquad (6.2.1)$$

The three regular diagrams D_+, D_-, D_0 formed as above are called *skein diagrams*, and the relation, (6.2.1), between the Laurent polynomials of K_+, K_-, K_0 (whose regular diagrams these are) is called the *skein relation*. Also, an operation that replaces one of D_+, D_-, D_0 by the other two is called a *skein operation*.

We shall write $\nabla_{D_+}(z)$ instead of $\nabla_{K_+}(z)$, *et cetera*, since there is no need to distinguish between the knots K_+, K_-, K_0 and their respective regular diagrams D_+, D_-, D_0.

Exercise 6.2.1. Show that if K_+ is a μ-component link, then K_- is also a μ-component link, but K_0 is either a $(\mu - 1)$-component or a $(\mu + 1)$-component link.

The polynomial $\nabla_K(z)$, defined as above, is called the *Conway polynomial*. To actually show that the Laurent polynomial $\nabla_K(z)$, obtained from Axioms 1 and 2, is well-defined and unique is quite troublesome (a complete proof can be found in [LM]). However, if we assume the well-definedness and uniqueness of $\nabla_K(z)$, then by proving the following theorem, we can show that $\nabla_K(z)$ and the Alexander polynomial are essentially the same.

Theorem 6.2.1.

$$\Delta_K(t) = \nabla_K\left(\sqrt{t} - \frac{1}{\sqrt{t}}\right)$$

In other words, if we replace z by $\sqrt{t} - \frac{1}{\sqrt{t}}$ in the Conway polynomial, the resultant transformation yields the Alexander polynomial. Due to this relationship, $\nabla_K(z)$ is often called the Alexander-Conway polynomial.

<u>Proof</u>

We have already inadvertently shown in Example 6.1.2 that the Alexander polynomial satisfies Axiom 1; therefore, we need only prove that it also satisfies Axiom 2, taking into account the substitution $z = \sqrt{t} - \frac{1}{\sqrt{t}}$.

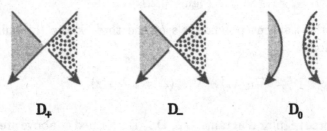

D_+ D_- D_0

Figure 6.2.2

Let us first consider the skein diagrams, Figure 6.2.2. If F_+, F_-, F_0 are the Seifert surfaces that correspond to D_+, D_-, D_0, and Γ_+, Γ_-, Γ_0 are the corresponding graphs, then we may, by using the methods of the previous chapter, determine the respective Seifert matrices, M_+, M_-, M_0.

The crossing point of D_+ (respectively D_-), see Figure 6.2.2, corresponds to a positive (negative) band in F_+ (respectively F_-), while in the graph Γ_+ (Γ_-) the crossing point corresponds to a positive (negative) edge e_+ (e_-), see Figure 6.2.3.

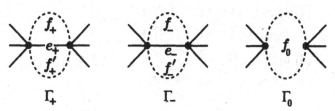

Figure 6.2.3

Let f_+ (f_-) be the domain of Γ_+ (Γ_-) that contains the edge e_+ (e_-). Suppose e_+ (e_-) is one of the common boundary edges of f_+ (f_-) and f'_+ (f'_-), Figure 6.2.3. It is possible that f_+ and f'_+ are the same face (and so are f_- and f'_-). In this case Γ_0 is disconnected and D_0 is not connected. Definition 5.4.2 and Proposition 5.4.2 then imply that $\Delta_{K_0}(t) = 0$ and $\nabla_{K_0}(z) = 0$. On the other hand, in this case K_+ and K_- are equivalent, since they are the connected sum of two knots (or links). Therefore, $\Delta_{K_+}(t) = \Delta_{K_-}(t)$, and hence $\Delta_K(t)$ satisfies Axiom 2. Therefore, we suppose that f_+ (f_-) and f'_+ (f'_-) are different. Now let the order of M_+ and M_- be k. We may then assume that f_+ (f_-) and f'_+ (f'_-) correspond to the $(k-1)^{\text{th}}$ row (and column) and the last k^{th} row (and column), respectively. Since f_+ (f_-) and f'_+ (f'_-) in Γ_+ (Γ_-) are "amalgamated" to form Γ_0, M_0 is of order $k-1$. Now, let $a_{i,j}(+)$, $a_{i,j}(-)$, $a_{i,j}(0)$ denote, respectively, the entries of M_+, M_-, M_0. Then these entries are related as follows, Figure 6.2.4:

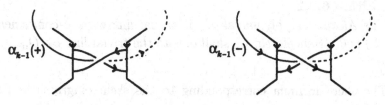

Figure 6.2.4

(A) If $i, j \neq k - 1, k$, then

$$(1) \quad a_{i,j}(+) = a_{i,j}(-) = a_{i,j}(0).$$

(B) If $i \neq k - 1$,

    (1) $a_{i,k-1}(0) = a_{i,k-1}(+) + a_{i,k-1}(-)$

    (2) $a_{k-1,i}(0) = a_{k-1,i}(+) + a_{k-1,i}(-)$.   (6.2.2)

(C) If $i = k - 1$, or k,

    (1) $a_{i,i}(+) = a_{i,i}(-) - 1$

    (2) $a_{k-1,k}(+) = a_{k-1,k}(-) + 1$

        $a_{k,k-1}(+) = a_{k,k-1}(-) + 1$

    (3) $a_{k-1,k-1}(0) = \displaystyle\sum_{i,j=k-1,k} \{a_{i,j}(+) + a_{i,j}(-)\}$.

Exercise 6.2.2. Show that (6.2.2) actually holds.

Using (6.2.2), a further simple calculation shows that

$$\det(M_+ - tM_+^T) - \det(M_- - tM_-^T) = (-1)(1 - t)\det(M_0 - tM_0^T).$$

So if we use this to calculate $\Delta_K(t)$, we obtain

$$\begin{aligned}
\Delta_{K_+}(t) &= t^{-\frac{k}{2}}\det(M_+ - tM_+^T) \\
&= t^{-\frac{k}{2}}\det(M_- - tM_-^T) + t^{-\frac{k}{2}}(t - 1)\det(M_0 - tM_0^T) \\
&= \Delta_{K_-}(t) + (t^{\frac{1}{2}} - t^{-\frac{1}{2}})\Delta_{K_0}(t).
\end{aligned}$$

This now completes the proof of Theorem 6.2.1.   ■

Before using the skein relation to calculate the Alexander polynomial, we shall prove the following proposition:

Proposition 6.2.2.

    The Alexander polynomial of the trivial link with μ-components ($\mu \geq 2$) is 0. (Henceforth, we shall denote the trivial link by O_μ.)

Proof

    The skein formula corresponding to the skein diagrams in Figure 6.2.5 is

$$\nabla_{D_+}(z) - \nabla_{D_-}(z) = z\nabla_{D_0}(z).$$

Since both D_+ and D_- are trivial ($\mu - 1$)-component links, $\nabla_{D_+}(z) = \nabla_{D_-}(z)$. Therefore, $0 = z\nabla_{D_0}(z)$, i.e., $\nabla_{D_0}(z) = 0$.

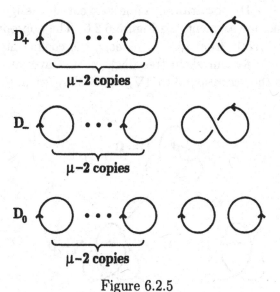

Figure 6.2.5

Proposition 6.2.2 is also an immediate consequence of Definition 5.4.2, since the Seifert matrix of O_μ contains the zero matrix of order $\mu - 1$.

Usually, the most effective way to calculate the Conway polynomial is to make use of the *skein tree diagram*. Since it is a calculating aid, it is best illustrated/defined by means of an example.

To facilitate our next set of calculations we shall rewrite (6.2.1) as follows:

$$\left.\begin{aligned} \nabla_{D_+}(z) &= \nabla_{D_-}(z) + z\nabla_{D_0}(z) \\ \nabla_{D_-}(z) &= \nabla_{D_+}(z) - z\nabla_{D_0}(z) \end{aligned}\right\} \tag{6.2.3}$$

Example 6.2.1.

Suppose that K is the right-hand trefoil knot and D is a regular diagram of K, which in Figure 6.2.6 is the topmost diagram.

Within the dotted circle on D, we will perform a skein operation. Since within this circle the crossing point is positive, it is better to rename this regular diagram D_+. By performing a skein operation, D_+ is transformed into two other regular diagrams: one, D_-, is the regular diagram obtained by changing the original positive crossing point to a negative crossing point; the other, D_0, is obtained by removing the positive crossing point (by splicing the regular diagram at this crossing point), see Figure 6.2.6. Now connect D_+ and D_- (and, similarly, connect D_+ and D_0) by drawing a line segment and assign +1 (respectively z) to the line segment, see Figure 6.2.6. The appropriate assignment follows directly from (6.2.3), namely, the coefficients of $\nabla_{D_-}(z)$ and $\nabla_{D_0}(z)$. So for our skein tree diagram we have our first pair of branches, and they correspond to $1\nabla_{D_-}(z) + z\nabla_{D_0}(z)$ in the evaluation of $\nabla_{D_+}(z)$.

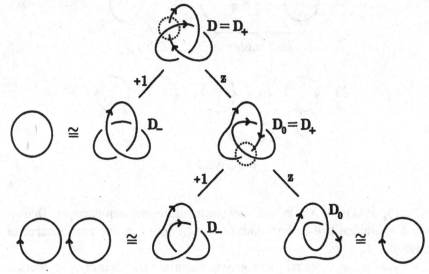

Figure 6.2.6

It is straightforward to see that D_- is equivalent to the trivial knot, and hence $\nabla_{D_-}(z) = 1$. Therefore D_- will not produce any further branches, i.e., we cannot perform a subsequent skein operation here. However, D_0 is not equivalent to the trivial knot or links, and so we can again perform a skein operation within another dotted circle. Since within this circle, see Figure 6.2.6, the crossing point is also positive, let us rename D_0 as D_+. Then, as before, let us denote the subsequent left-hand diagram by D_- and the right-hand diagram D_0. Again, we draw line segments and assign to them coefficients by means of (6.2.3), see Figure 6.2.6. It is easy to see that D_- and D_0 are, respectively, the trivial 2-component link and the trivial knot. Hence no further branches may be formed and our skein tree diagram for K is complete.

$\nabla_K(z)$ can now be calculated as the sum of the Conway polynomial of the each terminating trivial knot (or link) multiplied by the product of the coefficients (on the line segments) along the (uniquely determined) branch path that begins with our original regular diagram, D, of K and terminates, by construction, with the regular diagram of this trivial knot (or link).

Therefore, in the above example we obtain the following sum:

$$\nabla_K(z) = 1\nabla_O(z) + z\nabla_{OO}(z) + z^2\nabla_O(z).$$

Since $\nabla_O(z) = 1$ and $\nabla_{OO}(z) = 0$, the calculation collapses down to $\nabla_K(z) = 1 + z^2$. Therefore, by applying Theorem 6.2.1 we obtain

$$\Delta_K(t) = 1 + (\sqrt{t} - \frac{1}{\sqrt{t}})^2 = t^{-1} - 1 + t.$$

Example 6.2.2.

The skein tree diagram for the Conway polynomial for the figure 8 knot is given in Figure 6.2.7.

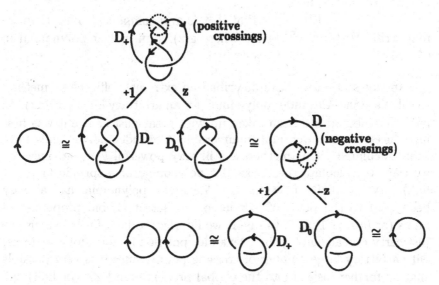

Figure 6.2.7

The following calculation is a direct result of this skein tree diagram:

$$\nabla_K(z) = \nabla_O(z) + z\nabla_{OO}(z) - z^2\nabla_O(z) = 1 - z^2.$$

Therefore,

$$\Delta_K(t) = 1 - (\sqrt{t} - \frac{1}{\sqrt{t}})^2 = -t^{-1} + 3 - t.$$

From the above two examples it is possible to surmise how in general we can calculate the Conway polynomial. We start with the regular diagram D of a knot (or link) and then perform on it an unknotting operation at one of the crossing points. This produces a new regular diagram D′ of a "simpler" knot. Continuing this process, we finally arrival at a set of regular diagrams of trivial knots and links. (The reader should consult Proposition 4.4.1 to see why, in fact, this is possible; however, a small *caveat*, in this case the number of crossing points of D_0 is one less than for D_+ and D_-.)

Exercise 6.2.3. Determine the Alexander and the Conway polynomials for the Borromean rings (Figure 1.4.5), and the Whitehead link (Figure 1.4.4) and the knot in Figure 5.1.5(a).

Exercise 6.2.4. (1) Show that if K is a knot, then $\nabla_K(z)$ is an integer polynomial in z^2.

(2) Show that if K is a μ-component link, then we may write $\nabla_K(z) = z^{\mu-1}g(z)$, where g(z) is an integer polynomial in z^2.

In this section we have described an extremely efficacious method to calculate the Alexander polynomial for an arbitrary knot (or link). In fact, this method has been taken up with great gusto since it was first introduced by Conway, and innumerable Alexander polynomials have been calculated. This method is still very powerful if in our research we want to calculate the Alexander polynomial of a specific knot (or link). By the 1960s, however, the Alexander polynomial had already been used to the point of exhaustion to detect Global properties of knots (or links). That is to say, it would seem to be a futile exercise to just carry on calculating the Alexander polynomial for knots (or links) with arbitrary large number of crossing points, since it is very possible that no further insight into the Global properties of knot (or link) will be garnered by so doing. Around 15 years after Conway introduced his method, it was shown that in addition to the Alexander polynomial, his approach is very useful in the calculation of the "new" knot invariants.

§3 Basic properties of the Alexander polynomial

In this section we shall prove several important properties of the Alexander polynomial. The first of these properties is related to Proposition 6.1.2.

Proposition 6.3.1.
   If K is a knot, then $\Delta_K(1) = 1$.

Proof
   We know, due to Theorem 6.2.1, that if we set $t = 1$ then $\Delta_K(1) = \nabla_K(0)$. On the other hand, by (6.2.1), if we set $z = 0$ then we obtain $\nabla_{D_+}(0) = \nabla_{D_-}(0)$. Therefore, even if we perform an unknotting operation to K, the value of $\nabla_K(0)$ remains unchanged. However, by performing the unknotting operation several times on K, eventually K is transformed into a trivial knot (Proposition 4.4.1). From the above observations we may deduce, $\nabla_K(0) = \nabla_O(0) = 1$.
                                                                          ∎

Exercise 6.3.1.   Show that if L is a μ-component $(\mu \geq 2)$ link, then $\Delta_L(1) = 0$.

   The reader may have noticed that if we juxtapose Proposition 6.3.1 and Theorem 6.1.4, then these actually seem to characterize the Alexander polynomial, as seen in Theorem 6.3.2.

Theorem 6.3.2.
   Suppose that f(t) *is a Laurent polynomial that satisfies the following two conditions:*

$$(1) \quad f(1) = 1$$
$$(2) \quad f(t) = f(t^{-1}). \tag{6.3.1}$$

Then there <u>exists</u> *a knot that has as its Alexander polynomial* f(t). *Equivalently, if* g(z) *is an integer polynomial in* z^2 *with* g(0) = 1, *then there exists a knot K that has as its Alexander-Conway polynomial* g(z).

   The proof requires finding an appropriate orientable surface, F, with its Seifert matrix M of order k satisfying $t^{-\frac{k}{2}}\det(M - tM^T) = f(t)$. The reader may wish to try and develop a proof or consult [Sei] for more details.

Example 6.3.1.   In Figure 6.3.1 we have drawn one of the knots, which has as its Alexander polynomial $f(t) = 2t^{-2} - 10t^{-1} + 17 - 10t + 2t^2$ $(= 1 - 2z^2 + 2z^4)$.

   The Alexander polynomial is particularly useful with regard to the Global problem of the classification of knots (or links). However, it is powerless in deciding the Local problems of invertibilty and amphicheirality.

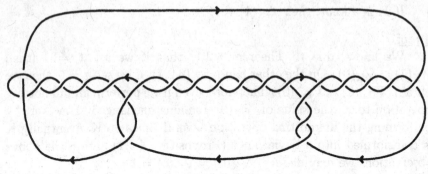

Figure 6.3.1

Theorem 6.3.3.

Suppose that K is a knot.

(1) If −K is the knot obtained from K by reversing the orientation on K, then

$$\Delta_K(t) = \Delta_{-K}(t).$$

(2) If K* is the mirror image of K, then

$$\Delta_{K^*}(t) = \Delta_K(t).$$

Proof

(1) From the proof of Proposition 5.4.6, we may assume that $M_{-K} = M_K^T$. If M_K is of order 2g, say, then M_{-K} is also of order 2g. So,

$$\Delta_K(t) = t^{-g} \det(M_K - tM_K^T)$$
$$= t^{-g} \det(M_{-K}^T - tM_{-K})$$
$$= t^{-g} \det(M_{-K} - tM_{-K}^T)^T = \Delta_{-K}(t).$$

(2) Since $M_{K^*} \overset{S}{\sim} -M_K^T$ and utilizing the proof of Proposition 5.4.7, we may assume that $M_{K^*} = -M_K^T$. This calculation now follows along similar lines to (1),

$$\Delta_{K^*}(t) = t^{-g} \det(M_{K^*} - tM_{K^*}^T)$$
$$= t^{-g} \det(-M_K^T + tM_K)$$
$$= (-1)^{2g} t^{-g} \det(M_K - tM_K^T)^T = \Delta_K(t).$$
∎

We know that the Alexander polynomial of the trivial knot is 1; the converse, however, does not hold, i.e., there exist non-trivial knots that have Alexander polynomial 1. An example of one such knot is given in Figure 6.3.2; this knot is known as the *Kinoshita-Terasaka* knot.

KT knot

Figure 6.3.2

The Alexander polynomial of the trivial μ-component $(\mu \geq 2)$ link has been shown to be 0. We may also generalize this result.

Definition 6.3.1. The regular diagram D of a link L that is a composition of the regular diagrams of two links with no points in common is said to be *split*, see Figure 6.3.3. A link L that has a split regular diagram is said to be a *split link*. (The link in Figure 6.3.3 is a 3-component split link.)

L

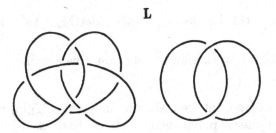

Figure 6.3.3

Proposition 6.3.4.
If L is a split link, then $\Delta_L(t) = 0$.

Proof
Suppose that D is a split regular diagram of L. Then the terminal point of the skein tree diagram of D is *always* a trivial link; it can never terminate in a trivial knot. (Why is this the case?) Therefore, it follows from Proposition 6.2.2 that $\nabla_L(z) = 0$, i.e., $\Delta_L(t) = 0$. ∎

Exercise 6.3.2. If $\Delta_L(t) = 0$ it does not necessarily follow that L is a split link. Show that the Alexander polynomial of the link in Figure 4.5.5(b) is 0.

Theorem 6.3.5.

Suppose $K_1 \# K_2$ *is the connected sum of two knots (or links)* K_1 *and* K_2, *then*

$$\Delta_{K_1 \# K_2}(t) = \Delta_{K_1} \Delta_{K_2}.$$

Proof

Firstly, create in the prescribed way the Seifert surfaces F_1 and F_2 of, respectively, K_1 and K_2. Then the orientable surface formed by joining these surfaces by a band becomes a Seifert surface for $K_1 \# K_2$, see Figure 2.1.7. If we suppose M_1 and M_2 are the Seifert matrices of K_1 and K_2 obtained from F_1 and F_2, then M the Seifert matrix of $K_1 \# K_2$ has the following form:

$$M = \begin{bmatrix} M_1 & 0 \\ 0 & M_2 \end{bmatrix}.$$

Therefore,

$$\det(M - tM^T) = \det(M_1 - tM_1{}^T)\det(M_2 - tM_2{}^T).$$

Theorem 6.3.5 follows immediately from this equality.

∎

If L is a μ-component link, then we may write $\nabla_L(z) = z^{\mu-1}g(z)$, where g(z) is an integer polynomial in z^2; cf. Exercise 6.2.4(2).

So, if we let

$$\tilde{\Delta}_L(t) = g(\sqrt{t} - \frac{1}{\sqrt{t}}),$$

then

$$\tilde{\Delta}_L(t^{-1}) = \tilde{\Delta}_L(t),$$

and thus $\tilde{\Delta}_L(t)$ is a symmetric integer polynomial. This polynomial is called the *Hosokawa* polynomial. With regard to the Hosokawa polynomial for links, the following theorem, similar to Theorem 6.3.2, holds:

Theorem 6.3.6 [H].

Suppose f(t), a Laurent polynomial, is symmetric, i.e., f(t) = f(t⁻¹). Then there exists a link, with an arbitrary number of components, that has as its Hosokawa polynomial f(t). [Note that f(1) can be an arbitrary integer.]

Exercise 6.3.3. Calculate and verify that the Hosokawa polynomial of the 2-component link L in Figure 6.3.4 is $\widetilde{\Delta}_L(t) = t^{-1}+3+t$ ($= 5+z^2$). Find a 3-component link L' with the same Hosokawa polynomial as L. (Hint: The Alexander-Conway polynomial of the Hopf link is z, and use Theorem 6.3.5.)

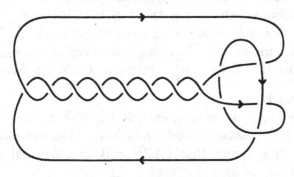

Figure 6.3.4

In general, the problem of determining the genus of a knot is quite difficult. However, it is possible to estimate, and for some cases to determine, the genus of a knot using the Alexander polynomial.

Suppose for a *knot*, K, of genus g(K), F is a Seifert surface for K with this particular genus. Then the order of the Seifert matrix M calculated from F is 2g(K). Therefore, the maximum degree of t in $\Delta_K(t)$ is at most g(K). We can put this result on a more formal footing in the following theorem:

Theorem 6.3.7.

Suppose that the genus of a knot K is g(K), then the maximum degree of t in the Alexander polynomial cannot exceed g(K).

In particular, if $\det M \neq 0$ then since $\det(M - tM^T)$ is a Laurent polynomial of exactly degree 2g, we obtain the next proposition.

Proposition 6.3.8.

A necessary and sufficient condition for the maximum degree of t in $\Delta_K(t)$, the Alexander polynomial of K, to be exactly g(K) is $\det M \neq 0$.

Using Proposition 6.3.8 we can prove that the genus of an alternating knot is equal to the maximal degree of its Alexander polynomial; cf. Chapter 11, Section 5.

§4 The signature of a knot

In this section, we shall define still one more important invariant that depends on the Seifert matrix M, the invariant in question is the signature of a knot. As was noted in Proposition 6.1.1, the absolute value of the determinant of the symmetric matrix $M + M^T$ is a knot invariant. However, this is not the only knot invariant that can be formed with this symmetric matrix as its core element.

A well-known result in linear algebra allows us to diagonalize any symmetric matrix that has in all its entries real numbers. More precisely,

Theorem 6.4.1.

Suppose A *is a* $n \times n$ *symmetric matrix with its entries real numbers. Then it is possible to find a real (with its entries real numbers) invertible matrix* P *such that* $PAP^T = B$ *is a diagonal matrix. In addition, we may assume that* $\det P = \pm 1$.

We may rephrase the essence of this theorem in the terminology of Chapter 5 as follows: A symmetric matrix is Λ_1-equivalent to a diagonal matrix.

Instead of giving a proof of Theorem 6.4.1, which is a bit tedious and will not shed any insight in what follows, we propose to illustrate the method of diagonalizing a matrix by means of a couple of examples. These examples will, hopefully, indicate to the reader the idea of the proof, and thus the proof will become only an exercise in (the manipulations of) linear algebra.

Example 6.4.1.

(a) Let us diagonalize the matrix,

$$A = \begin{bmatrix} 0 & 1 & -1 \\ 1 & 2 & 1 \\ -1 & 1 & 3 \end{bmatrix}.$$

First interchange the first row and the second row, and then also interchange the first and second columns. By this two manipulations, the new matrix A_1 will not have 0 in its $(1,1)$-entry:

$$A \rightarrow \begin{bmatrix} 2 & 1 & 1 \\ 1 & 0 & -1 \\ 1 & -1 & 3 \end{bmatrix} = A_1.$$

Next, we will apply several Λ_1-operations to eliminate all the entries in the first row and column except the $(1,1)$-entry.

$$A_1 \rightarrow \begin{bmatrix} 2 & 0 & 0 \\ 0 & -\frac{1}{2} & -\frac{3}{2} \\ 0 & -\frac{3}{2} & \frac{5}{2} \end{bmatrix} = A_2.$$

Next, consider the 2×2 matrix, with non-zero elements, which forms the lower right-hand block of A_2. On this 2×2 matrix apply, in a similar way to our previous step, several Λ_1-operations, so that within this 2×2 matrix, also, all the entries in the first row and column except the $(1,1)$-entry are eliminated (in fact, we need only eliminate two entries):

$$A_2 \rightarrow \begin{bmatrix} 2 & 0 & 0 \\ 0 & -\frac{1}{2} & 0 \\ 0 & 0 & 7 \end{bmatrix} = B.$$

(b) Let us next diagonalize:

$$A = \begin{bmatrix} 0 & 1 \\ 1 & 0 \end{bmatrix}.$$

Since all the diagonal entries are 0, we cannot apply the previous method. However, by adding the second row to the first row, and then adding the second column to the first column, we can get a new matrix, A_1, with non-zero $(1,1)$-entry:

$$A \rightarrow \begin{bmatrix} 2 & 1 \\ 1 & 0 \end{bmatrix} = A_1.$$

We may now diagonalize A_1 as in (a), and so

$$A_1 \rightarrow \begin{bmatrix} 2 & 0 \\ 0 & -\frac{1}{2} \end{bmatrix} = B.$$

Returning, now, to Theorem 6.4.1, suppose that the entries in the diagonal of B are $a_1, a_2, \ldots, a_n$. We may lift the following two, seemingly innocuous, definitions directly from their linear algebra roots.

(1) The number of $a_1, a_2, \ldots, a_n$ that are zero

    is called the *nullity* of A, and is denoted by n(A).   (6.4.1)

(2) (The number of positive a_i) $-$ (the number of negative a_i)

    is called the *signature* of A, and is denoted by $\sigma(A)$.

The signature, $\sigma(A)$, and the nullity, n(A), depend, as the notation implies, only on A, completely irrespective of the P chosen. The signature may also, although this does not follow directly from the above definition, be obtained from the following:

$$\sigma(A) = \text{(the number of positive eigenvalues of A)} - $$

$$\text{(the number of negative eigenvalues of A)}.$$

Exercise 6.4.1.

(a) Show that if (at least) one of the diagonal entries of a $n \times n$ symmetric matrix is positive, then $\sigma(A) > -n$.

(b) Diagonalize A, given below, and thereupon calculate $\sigma(A)$:

$$A = \begin{bmatrix} 2 & 1 & 2 & 0 \\ 1 & 0 & 3 & -1 \\ 2 & 3 & 2 & 1 \\ 0 & -1 & 1 & -2 \end{bmatrix}.$$

Our intention is to prove that the signature of $M + M^T$ is, in fact, a knot invariant, but first we need to prove a general theorem.

Theorem 6.4.2.

If N_1 and N_2 *are two integer square matrices that are* S-equivalent *then*

$$n(N_1 + N_1^T) = n(N_2 + N_2^T)$$

$$\sigma(N_1 + N_1^T) = \sigma(N_2 + N_2^T).$$

Proof

To avoid getting tangled up in notation, let $N = N_1 + N_1^T$ and $N' = N_2 + N_2^T$. In order to prove the theorem, we need to show the following two properties:

(1) If $N_2 = \Lambda_1(N_1)$, then $n(N) = n(N')$ and $\sigma(N) = \sigma(N')$

(2) If $N_2 = \Lambda_2(N_1)$, then $n(N) = n(N')$ and $\sigma(N) = \sigma(N')$.   (6.4.2)

Proof of (1)

If $N_2 = \Lambda_1(N_1)$, then we can write $N_2 = PN_1P^T$. Therefore,

$$N' = PN_1P^T + (PN_1P^T)^T = PNP^T.$$

In addition to this, we know that P is an invertible matrix. Using these facts, the required results, namely, $n(N') = n(N)$ and $\sigma(N') = \sigma(N)$, follow directly from well-known results in linear algebra.

Proof of (2)

If $N_2 = \Lambda_2(N_1)$, then we can write

$$N_2 = \left[\begin{array}{ccccc} & & & b_1 & 0 \\ & N_1 & & \vdots & \vdots \\ & & & b_m & 0 \\ 0 & \cdots & 0 & 0 & 1 \\ 0 & \cdots & 0 & 0 & 0 \end{array} \right] \; \left(\text{or} \; \left[\begin{array}{ccccc} & & & 0 & 0 \\ & N_1 & & \vdots & \vdots \\ & & & 0 & 0 \\ b_1 & \cdots & b_m & 0 & 0 \\ 0 & \cdots & 0 & 1 & 0 \end{array} \right] \right).$$

So,

$$N' = N_2 + N_2^T = \left[\begin{array}{ccccc} & & & b_1 & 0 \\ & N & & \vdots & \vdots \\ & & & b_m & 0 \\ b_1 & \cdots & b_m & 0 & 1 \\ 0 & \cdots & 0 & 1 & 0 \end{array} \right].$$

Now, if we choose our invertible matrix P judiciously (we leave this as a straightforward exercise for the reader), we can obtain

$$PN'P^T = \left[\begin{array}{cc} N & O \\ O & \begin{array}{cc} 0 & 1 \\ 1 & 0 \end{array} \end{array} \right].$$

Therefore,

$$n(N') = n(N) + n\left[\begin{array}{cc} 0 & 1 \\ 1 & 0 \end{array} \right], \quad \sigma(N') = \sigma(N) + \sigma\left[\begin{array}{cc} 0 & 1 \\ 1 & 0 \end{array} \right].$$

However,

$$\det\left[\begin{array}{cc} 0 & 1 \\ 1 & 0 \end{array} \right] = -1 \neq 0, \quad \text{so} \quad n\left[\begin{array}{cc} 0 & 1 \\ 1 & 0 \end{array} \right] = 0,$$

and hence $n(N') = n(N)$.

In the case of the signature, we have already shown in Example 6.4.1(b) that we may diagonalize $\begin{bmatrix} 0 & 1 \\ 1 & 0 \end{bmatrix}$ as $\begin{bmatrix} 2 & 0 \\ 0 & -\frac{1}{2} \end{bmatrix}$.

It follows immediately from this that $\sigma \begin{bmatrix} 0 & 1 \\ 1 & 0 \end{bmatrix} = 0$. Therefore, also, $\sigma(N') = \sigma(N)$. ∎

Theorem 6.4.3.

Suppose M is the Seifert matrix of a knot (or link) K. If we set $A = M + M^T$, then $n(A)$, the nullity of A, and $\sigma(A)$, the signature of A, are invariants of the knot (or link) K. Hence, we can write $\sigma(K)$ instead of $\sigma(A)$, and this is called the signature of the knot (or link) K. Similarly, we can write $n(K)$ instead of $n(A)$, and this is called the nullity of K.

Proof

Since the two matrices M_1 and M_2 of K are S-equivalent, it follows from Theorem 6.4.2 that the signature and nullity of $M_1 + M_1^T$ and $M_2 + M_2^T$ are equal. The theorem follows directly from this result. ∎

Theorem 6.4.4.

If K is a knot, then $n(K) = 0$ and $\sigma(K)$ is always even.

Proof

The Seifert matrix, M, for K is a square matrix of even order. Further, due to Proposition 6.3.1, since $\det(M - M^T) = \Delta_K(1) = 1$, $\det(M + M^T)$ is an odd integer and so non-zero. Consequently, $n(M + M^T) = 0$, and $n(K) = 0$. Therefore, the number of eigenvalues of $M + M^T$ that are not zero is even; hence, $\sigma(M + M^T)$ is also even. ∎

Example 6.4.2. The Seifert matrix, M, of the right-hand trefoil knot, calculated in Example 5.3.2, is

$$\begin{bmatrix} -1 & 0 \\ 1 & -1 \end{bmatrix}.$$

Therefore,

$$M + M^T = \begin{bmatrix} -2 & 1 \\ 1 & -2 \end{bmatrix}.$$

We may diagonalize this in the same way as our previous examples to obtain

$$\begin{bmatrix} -2 & 0 \\ 0 & -\frac{3}{2} \end{bmatrix};$$

so $\sigma(K) = -2$. Similarly, the signature of the left-hand trefoil knot, K', may be shown to be 2.

Calculating the signature is a very effective way of rooting out the local properties of a knot, and hence it has had a profound influence in the solution of Local problems. As probably the most obvious example of this, we shall look at the problem of amphicheirality. To distinguish a knot K from its mirror image, K^*, is, in general, a very difficult problem. To show that the trefoil knot and its mirror image are not equivalent originally required quite an elaborate proof by Dehn. However, this follows immediately from the next theorem.

Theorem 6.4.5.

The signature of a knot (or link) K has the following properties:

(1) $\sigma(K_1 \# K_2) = \sigma(K_1) + \sigma(K_2)$.

(2) *If K^* is the mirror image of K, then $\sigma(K^*) = -\sigma(K)$.* (6.4.3)

(3) *If $-K$ is obtained from K by reversing the orientation on K, then $\sigma(-K) = \sigma(K)$.*

Proof

(1) It was shown in the proof of Theorem 6.3.5 that $M_{K_1 \# K_2}$ has the following form:

$$\begin{bmatrix} M_{K_1} & O \\ O & M_{K_2} \end{bmatrix}.$$

So,

$$\sigma(K_1 \# K_2) = \sigma(M_{K_1 \# K_2} + M_{K_1 \# K_2}^T)$$
$$= \sigma(M_{K_1} + M_{K_1}^T) + \sigma(M_{K_2} + M_{K_2}^T)$$
$$= \sigma(K_1) + \sigma(K_2).$$

(2) From Proposition 5.4.7 we know that M_{K^*} and $-M_K^T$ are S-equivalent, so by Theorem 6.4.2 we have

$$\sigma(K^*) = \sigma(M_{K^*} + M_{K^*}^T)$$
$$= \sigma(-M_K^T - M_K)$$
$$= (-1)\sigma(M_K + M_K^T)$$
$$= -\sigma(K).$$

(3) We leave the proof of this part as an exercise for the reader.
∎

Proposition 6.4.6.
If a knot (or link) K is amphicheiral then $\sigma(K) = 0$.

Proof
If K is amphicheiral then K and K^* are equivalent. Consequently, $\sigma(K) = \sigma(K^*)$. On the other hand, from (2) of (6.4.3) $\sigma(K^*) = -\sigma(K)$. So, $\sigma(K) = \sigma(K^*) = -\sigma(K)$. Hence, $\sigma(K) = 0$.
∎

Example 6.4.3. If K is the right-hand trefoil knot, we have shown $\sigma(K) = -2$. Consequently, K cannot be amphicheiral.

Example 6.4.4. The signature of the square knot [Figure 1.5.6(a)] is zero, while the signature of the granny knot [Figure 1.5.6(b)] is -4. So, these two knots are not equivalent (cf. Exercise 1.5.2). It can be shown that the square knot is, in fact, amphicheiral.

Exercise 6.4.2. Let $\Delta_K(t)$ be the Alexander polynomial of a knot (*but* not a link) K. Show

$$\text{sign}(\Delta_K(-1)) = (-1)^{\frac{\sigma(K)}{2}}, \tag{6.4.4}$$

where $\text{sign}(\Delta_K(-1))$ is just the sign of $\Delta_K(-1)$, i.e., either $+1$ or -1.

Using (6.4.4), show

$$|\Delta_K(-1)| \equiv (-1)^{\frac{\sigma(K)}{2}} \pmod 4. \tag{6.4.5}$$

Show the above formulae hold in practice in the case of the trefoil knot and the square knot.

It follows from (6.4.5) that a knot with $|\Delta_K(-1)| \equiv -1 \pmod 4$ cannot be amphicheiral. A more detailed discussion of the relationship between $\Delta_K(t)$ and $\sigma(K)$ can be found in a paper by J. Milnor [Mi]. In particular, he proves that if $\Delta_K(t) = 1$ then $\sigma(K) = 0$.

To calculate $\sigma(K)$ using the methods so far described requires the calculation of the Seifert matrix. However, we may avoid this by using the skein formula.

Theorem 6.4.7 [G].

Suppose K is a knot (but not a link) and D is a regular diagram for K. Then σ(K) can be determined by means of the following three axioms.

   (1) *If K is the trivial knot, then σ(K) = 0.*

   (2) *If* D_+, D_-, D_0 *are the skein diagrams, then*

$$\sigma(D_-) - 2 \le \sigma(D_+) \le \sigma(D_-) \tag{6.4.6}$$

   (3) *If* $\Delta_K(t)$ *is the Alexander polynomial of K, then*

$$\text{sign}(\Delta_K(-1)) = (-1)^{\frac{\sigma(K)}{2}}.$$

Since we will not give a proof of the above theorem, we would like to make a few short comments on (2) and (3) of (6.4.6). Firstly, because K is a knot (but not a link), D_+ and D_- are regular diagrams of knots. Since D_0 is a link, $\sigma(D_0)$ cannot be calculated from these two regular diagrams. Moreover, from Theorem 6.4.4, we know that $\sigma(K)$ is always even, which means that $\sigma(D_+)$ cannot be $\sigma(D_-) - 1$. Therefore, we may rewrite (2) of (6.4.6) as

$$(2)' \quad \sigma(D_+) = \sigma(D_-) \text{ or } \sigma(D_+) = \sigma(D_-) - 2.$$

Further, (6.4.6) part (3) is exactly (6.4.4). Inconveniently, there are no known similar skein formulae that allow us to evaluate the signature of a link.

Let us illustrate how Theorem 6.4.7 may be used to calculate the signature of the figure 8 knot.

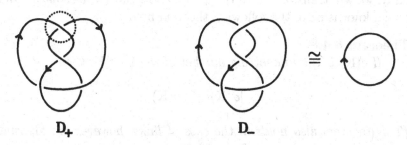

D₊   **D₋**

Figure 6.4.1

Example 6.4.5. If we consider the skein diagrams in Figure 6.4.1, it is easy to see that D_- is the trivial knot, so from $(2)'$ of (6.4.6),

$$\sigma(D_+) = \sigma(D_-) = 0 \quad \text{or} \quad \sigma(D_+) = -2.$$

On the other hand, since $\Delta_K(t) = -t^{-1}+3-t$ (cf. Example 6.2.2),

$$\Delta_K(-1)) = 5 \quad \text{so} \quad \text{sign}(\Delta_K(-1)) = 1.$$

We may now substitute this positive value into (3) of (6.4.4), and so obtain the equality $1 = (-1)^{\frac{\sigma(K)}{2}}$. It then follows immediately that $\sigma(K)$ must be zero, rather than -2. Since it is known that the figure 8 knot is amphicheiral, this is the expected result.

Exercise 6.4.3. Calculate the signature of the right-hand and left-hand trefoil knots using Theorem 6.4.7.

Exercise 6.4.4. Confirm that the signature in Figure 6.4.2 is zero. This knot, however, is known *not* to be amphicheiral.

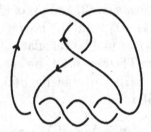

Figure 6.4.2

If we perform a single unknotting operation on a knot K, then we change D_+ to D_- or D_- to D_+. It follows from (2) of (6.4.6) that the signature is unchanged or changes only by ± 2. If u(K) is the unknotting number of K, then by performing the unknotting operations u(K) times on K, we will transform K to the trivial knot. Since the signature of the trivial knot is zero, the following theorem holds:

Theorem 6.4.8.

If u(K) is the unknotting number of the knot K, then

$$|\sigma(K)| \leq 2u(K).$$

(This theorem also holds in the case of links; however, an alternative proof to the knot case is required.)

Example 6.4.6. The signature of the knot, K, in Figure 6.4.3 is -4. So, $|\sigma(K)| = 4 \leq 2u(K)$, and hence $u(K) \geq 2$. However, only two knotting operations are required to transform K to the trivial knot; hence, $u(K) \leq 2$. From these two facts it follows that $u(K) = 2$.

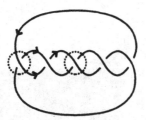

Figure 6.4.3

Sadly, Theorem 6.4.8 is not sufficient if we wish to determine the unknotting number of an arbitrary knot. Except for particular classes of knots, see Chapter 7, a general method for determining u(K) is not known.

However, $\sigma(K)$ is one of the few knot invariants that gives a necessary condition for a knot to be a slice knot.

Theorem 6.4.9 [Mus3].
If K is a slice knot, then $\sigma(K) = 0$.

It is also possible to use the Alexander polynomial to obtain conditions for a knot to be a slice knot [FM].

Exercise 6.4.5.   Calculate the signature of the knot in Figure 6.1.1(a) and determine u(K) for it. Do the same for the knot in Figure 6.1.1(b).

Exercise 6.4.6.   If L is a μ-component link show that $n(L) \le \mu - 1$.

Exercise 6.4.7.   Let L be an oriented μ-component $(\mu \ge 2)$ link. Let L' be the link obtained from L by reversing the orientation of one, say, μ^{th} component. Then it is known [Mus4] that

$$\sigma(L) + \text{lk}(L) = \sigma(L') + \text{lk}(L'), \qquad (6.4.7)$$

where lk(L) is the total linking number of L (for the definition see Chapter 4, Section 5). Therefore, $\xi(L) = \sigma(L) + \text{lk}(L)$ does not depend on the orientation of L, i.e., $\xi(L)$ is an invariant of an unoriented link L. *Nota bene,* $\sigma(L)$ may change considerably when the orientation of some of the components are reversed, since $\sigma(L)$ depends on the Seifert surface. Finally, the problem is as follows: Confirm that (6.4.7) holds for the link in Figure 4.5.5(a).

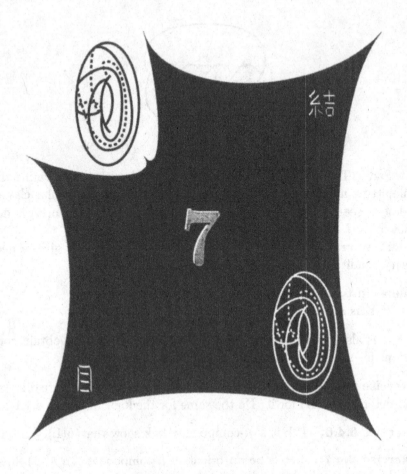

Torus Knots

If we take two knots (or links) at random, what we would like to have is an efficient method that will determine for us whether or not they are equivalent knots (or links). In general, sadly, such an efficient method has yet to be discovered. So, at present a concise classification of knots is not possible. The next most obvious step is to try to group together knots (or links) with a particular property or properties in common, and then try to classify them. In fact, the techniques we have already discussed are sufficient for us to extract the characteristics of certain particular types of knots.

In this chapter, using our aforementioned techniques, we shall investigate torus knots, which form one such set of knots (and links) with certain properties in common. The torus knots are not only interesting in themselves, but have a further importance in that it is often possible to gain insight into the general properties of knots (or links) by extrapolating the characteristics of the torus knots.

Although torus knots can be completely *classified* by our already established methods, we should add a *caveat* that these methods are not sufficient to totally determine all the properties of torus knots. For example, it is only quite recently that the general formula for the unknotting number of torus knots has been established. This case shows what we have already mentioned that the solution of Local problems does not follow automatically from the solution of the Global problem.

§1 Torus knots

A knot (or link) is a torus knot if it is equivalent to a knot (or link) that can be drawn without any points of intersection on the trivial torus. The trivial torus is a solid T obtained by rotating around the y-axis the circle $m : (x - 2)^2 + y^2 = 1$, on the xy-plane, which has as its centre the point (2,0), radius 1 unit, Figure 7.1.1.

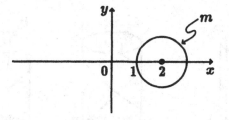

Figure 7.1.1

An alternative way to construct the trivial torus is to take a cylinder in $\mathbf{R}^3$ with the unit circle C_1 as its base and the unit circle C_2 as its top, Figure 7.1.2(a). We now glue together C_1 and C_2 in $\mathbf{R}^3$ so that C, the central axis of the cylinder, becomes the trivial knot, Figure 7.1.2(b). [*Nota bene*, if we glue C_1 and C_2 in the way shown in Figure 7.1.2(c), then the subsequent C is not equivalent to the trivial knot and hence the torus is not the trivial torus.]

A knot (or link) that lies on the trivial torus is said to be a torus knot, and we can express it in terms of a certain standard form.

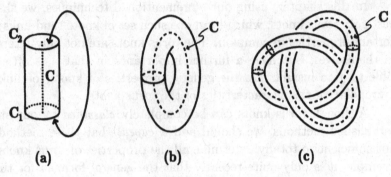

Figure 7.1.2

Let us use more concretely the above cylinder with height 1 unit and as its base a unit circle in xy-plane. We may assign to the base C_1 and to the top C_2 the r points $A_0, A_1, \ldots, A_{r-1}$ and $B_0, B_1, \ldots, B_{r-1}$, respectively. The co-ordinates are as follows, see also Figure 7.1.3(a):

$$A_0 = (1,0,0), \quad A_1 = (\cos \tfrac{2\pi}{r}, \sin \tfrac{2\pi}{r}, 0), \ldots,$$
$$A_{r-1} = (\cos \tfrac{2(r-1)\pi}{r}, \sin \tfrac{2(r-1)\pi}{r}, 0)$$
$$B_0 = (1,0,1), \quad B_1 = (\cos \tfrac{2\pi}{r}, \sin \tfrac{2\pi}{r}, 1), \ldots,$$
$$B_{r-1} = (\cos \tfrac{2(r-1)\pi}{r}, \sin \tfrac{2(r-1)\pi}{r}, 1).$$

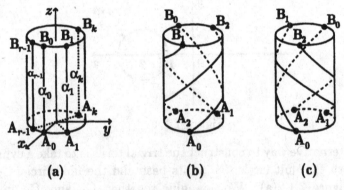

Figure 7.1.3

Let us now connect the point A_k and B_k $(k = 0, 1, \ldots, r-1)$ on the cylinder by the segments α_k. Next, keeping the base, C_1, fixed, let us give the whole cylinder a twist by rotating the top about the z-axis by an angle of $\frac{2\pi q}{r}$. (In this case, q is either a positive or negative integer.)

In Figure 7.1.3(b) the case $q = 2$ and $r = 3$ is shown, while in Figure 7.1.3(c) the case $q = -2$ and $r = 3$ is shown.

Finally, let us identify (i.e., glue in a very natural way) the point $(x, y, 0)$ of C_1 to the point $(x, y, 1)$ of C_2 (as before, the centre C becomes the trivial knot). This creates a single trivial torus T, with the r segments $\alpha_0, \alpha_1, \ldots, \alpha_{r-1}$ that have been transformed into a knot (or link) on its surface. This knot (or link) is called a (q, r)-*torus knot (or link)* and is denoted by $K_{q,r}$, Figure 7.1.4.

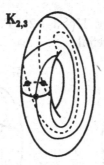

Figure 7.1.4

Let us once again consider the circle m that we used originally to form the trivial torus T. Further, suppose K is the boundary of a disk that lies on T, Figure 7.1.5.

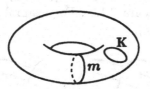

Figure 7.1.5

These trivial knots m and K, or a trivial link consisting of some of these trivial knots on T, is *not* a torus knot as outlined above. In fact, we shall consider these type of knots (or links) to correspond to the case $r = 0$, and we say that m is a $(1, 0)$-torus knot, $K_{1,0}$. The other knot, K, we shall call a $(0, 0)$-torus knot, $K_{0,0}$.

Exercise 7.1.1. Suppose that n is the greatest common divisor (g.c.d) of q and r. Show that $K_{q,r}$ is a n-component link.

Exercise 7.1.2. Given a torus knot (or link), that does not contain within its components the $(1, 0)$-torus knot and/or $(0, 0)$-torus knot (also excluding orientation), show that it is equivalent to some (q,r)-torus knot (or link).

In the above exercise we excluded orientation. However, it is a straightforward matter to assign an orientation to a torus knot $K_{q,r}$; just assign an orientation to each segment α_i, the orientation should flow from A_i to B_i. We shall denote this oriented torus knot $K(q,r)$. In addition, if we now reverse the orientation of each α_i, then we shall denote the subsequent oriented torus knot $K(-q,-r)$.

Example 7.1.1. $K(3,2)$ and $K(-3,-2)$, Figure 7.1.6, are equivalent to the versatile right-hand trefoil.

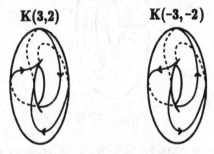

Figure 7.1.6

Exercise 7.1.3. Show that $K(q,r)$, $q, r > 0$, has a regular diagram of the form shown in Figure 7.1.7. For example, Figure 5.3.6(a) is a regular diagram for $K(3,4)$.

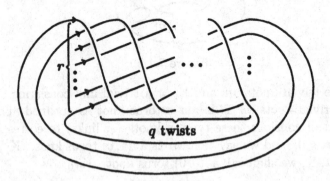

Figure 7.1.7

Exercise 7.1.4. Draw the oriented regular diagrams, of the form in Figure 7.1.7, for the torus knot $K(q,r)$ in the following three cases: (i) $q > 0, r < 0$; (ii) $q < 0, r > 0$; (iii) $q < 0, r < 0$.

We know that from a torus knot $K_{q,r}$ we can obtain either the (oriented) torus knot $K(q,r)$ or $K(-q,-r)$. However, it is also possible to obtain from the torus link $K_{q,r}$ an oriented torus link that is neither of these. For example, suppose $q = 4$ and $r = 2$, and α_0 has an upward orientation, while assigning to α_1 a downward orientation, Figure 7.1.8(a).

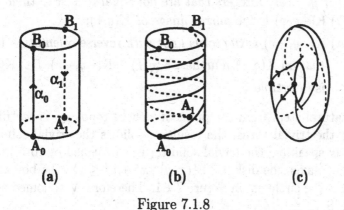

(a)   (b)   (c)

Figure 7.1.8

Then, by identifying the ends as described above, we shall obtain an oriented torus link, Figure 7.1.8(b) and (c). However, this torus link is neither K(4,2) nor K(-4,-2). Further, there do not exist a q and r such that it will become an oriented torus link of the type $K(q,r)$, cf. Exercise 7.3.3. Therefore to avoid such confusion occurring, we shall consider, from this point onwards, only *oriented* torus knots or links $K(q,r)$, with the orientation assigned as described above.

Exercise 7.1.5.   Show that the knot in Figure 6.1.1(a) is a torus knot and determine its type, i.e., find q and r.

Exercise 7.1.6.   Show that each component of a torus link $K(q,r)$ is a torus knot, and determine its type.

§2 The classification of torus knots (I)

In this and the following sections, to avoid convoluted notation, we shall mostly consider torus knots rather than torus links. Most of the results, however, can be shown to also hold for torus links, see Exercise 7.4.1.

If we draw the diagrams that correspond to the torus knots in the following proposition, then the results given in the proposition come out almost immediately.

Proposition 7.2.1.

Suppose $g.c.d(q,r) = 1$ and $r \neq 0$.

(1) If $q = 0, \pm 1$ or $r = \pm 1$, then $K(q,r)$ is the trivial knot.

(2) If q, r are integers that are not equal to $0, \pm 1$, then

   (i) $K(-q,r)$ is the mirror image of $K(q,r)$;

   (ii) $K(-q,-r)$ is the torus knot with reverse orientation to that on $K(q,r)$. Further, $K(q,r) \cong K(-q,-r)$, i.e., $K(q,r)$ is invertible.

A natural step, from the point of view of topology, is to "fill" the inside of the trivial torus, the resulting solid is the trivial solid torus V. Strictly speaking, the trivial solid torus is obtained by rotating once round the y-axis the disk $D^2 : \{(x,y) \mid x^2 + y^2 \leq 1\}$; the boundary of this disk is the circle m in Figure 7.1.1. Therefore, V is homeomorphic to $D^2 \times S^1$.

Proposition 7.2.2.

Suppose that V is a trivial solid torus in S^3. If V^o is the set of all the <u>internal</u> points of V, then $S^3 - V^o$ is also a trivial solid torus in S^3.

<u>Proof</u>

In the same way as explained in Chapter 1, Section 5, we shall consider S^3 to be constructed from two 3-balls that have been glued along their respective boundaries, namely, 2-spheres.

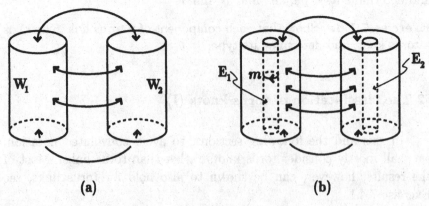

(a)   (b)

Figure 7.2.1

We may alter these two 3-balls so that they become two solid cylinders, W_1 and W_2. Let us now glue these cylinders together along their surfaces. First, we glue the top, base, and side surface of W_1 to the respective top, base, and side surface of W_2, Figure 7.2.1(a).

Next, consider relatively small, in terms of diameter, cylinders E_1 and E_2 from W_1 and W_2, respectively, Figure 7.2.1(b). These two cylinders may be thought of as the "fattened" centres of W_1 and W_2. If we now glue W_1 and W_2 along their boundaries in the above manner, then E_1 and E_2 form a trivial solid torus, V, in S^3 $(= W_1 \cup W_2)$. Therefore, to prove Proposition 7.2.2, it is sufficient to show that the solid obtained by extracting the internal points of this V from S^3 is also a solid torus.

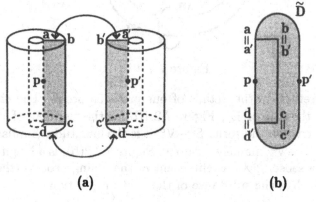

(a)　　　　　　　　　　**(b)**

Figure 7.2.2

As in Figure 7.2.2, in each of W_1 and W_2 create a rectangle. These two rectangles form a disk $\tilde{D}$ in S^3 when we glue together the surfaces of W_1 and W_2. To be precise, we glue together the edges ab and a'b', bc and b'c', and cd and c'd', Figure 7.2.2(b). If we now rotate the disk once round E_1 and E_2, which have also been glued together in the prescribed manner, then what we obtain is $S^3 - V^\circ$, and by construction this is also a (trivial) solid torus.
　　　　　　　　　　　　　　　　　　　　　　　　■

Contained in the above proof is something that in itself is of importance. In order to explain it clearly, let us first consider the two special circles on the trivial torus T that were discussed briefly in the previous section. As before, suppose that m is the circle from which we obtain T. By rotating the point P(3,0) on m once around the y-axis, we shall form another circle l. We say that m is the *meridian* of T and l is the *longitude* of T. (In general, the meridian and longitude are circles on T that are obtained by performing elementary knot moves on T to m and l, respectively.) Moreover, we say that the circle C, obtained by rotating the centre, (2,0), of the circle m is the centre of T. We may think of C and l as being "parallel." Since C and m form a link in $\mathbf{R}^3$, we may assign an orientation to C and m such that the linking number $\mathrm{lk}(m, C) = 1$, Figure 7.2.3. In addition, we assign the same orientation to l as on C. (We shall also say that m and l are the meridian and longitude of the trivial solid torus V.)

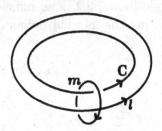

Figure 7.2.3

Let us return to the details of our previous proof. The circle m surrounding the core of E_1, Figure 7.2.1(b), is the meridian of V. However, for the trivial solid torus $S^3 - V^o$ it is the longitude. Similarly, the meridian of $S^3 - V^o$, namely, apdp$'$a, Figure 7.2.2(b), is a longitude for V. This, in a sense, gives a refinement of the gluing process, the exact nature of which is the substance of the next proposition.

Proposition 7.2.3.

S^3 *can be obtained by identifying (gluing) the surfaces (i.e., tori) of two trivial solid tori* V_1 *and* V_2 *in such a way that the meridian and longitude of* V_1 *glue to the longitude and meridian, respectively, of* V_2.

This proposition can be put to immediate use in proving the following theorem:

Theorem 7.2.4.

$$K(q, r) \cong K(r, q).$$

Proof

By Proposition 7.2.3, we may think of S^3 as two trivial solid tori that have been glued together along their surfaces, the (trivial) tori T_1 and T_2, respectively. Further, the meridian and longitude of T_1

correspond to the longitude and meridian of T_2. Finally, in our assumptions, we may think of $K(q,r)$ as a (q,r)-torus knot that lies on T_1 (cf. Exercise 7.2.1 below). However, if instead we think of this knot to be on T_2, then due to the interchange of meridians and longitudes it is a (r,q)-torus knot. But the knot itself has not been altered in any way; hence, $K(q,r) \cong K(r,q)$.

                                                                                    ∎

Exercise 7.2.1.   Show that a torus knot (or link) $K(q,r)$ is equivalent to a knot (or link) K on a trivial torus T such that K intersects a meridian, m, of T at exactly r points and a longitude, l, of T at $|q|$ points.

The above proof is a mathematical sleight of hand, but the above theorem will become clearer if instead we use the Reidemeister moves to transform the regular diagram of $K(q,r)$ into the regular diagram of $K(r,q)$.

Exercise 7.2.2.   Show using the Reidemeister moves that the regular diagrams, with orientation assigned, of $K(5,3)$ and $K(3,5)$ are equivalent.

Exercise 7.2.3.   Find a formula that determines the total linking number of the torus link $K(q,r)$.

Exercise 7.2.4.   Show that the Whitehead link and the Borromean rings are not torus links.

§3 The Seifert matrix of a torus knot

Before we classify the torus knots $K(q,r)$, we would like to look at their Seifert matrices. Suppose, firstly, that both q and r are positive integers. The knot in Figure 5.3.6(a), whose Seifert matrix we have already investigated in detail in Example 5.3.4, is in fact the torus knot $K(3,4)$. It is reasonable to infer from this example that the Seifert matrix of $K(q,r)$ has the following form.

Proposition 7.3.1.

Let q and r be positive integers. The Seifert matrix, M, of $K(q,r)$ is (S-equivalent to) a square matrix of order $(r-1)(q-1)$, which can be divided into the following $(r-1)^2$ blocks:

$$M = \begin{bmatrix} B_{11} & & & & & \\ B_{21} & B_{22} & & & O & \\ & B_{32} & B_{33} & & & \\ & & & \ddots & \ddots & \\ O & & & \ddots & B_{r-2,r-2} & \\ & & & & B_{r-1,r-2} & B_{r-1,r-1} \end{bmatrix},$$

such that
   (1) Each block is a square matrix of order $(q-1)$;
   (2) The diagonal blocks $B_{i,i}$ $(i = 1, 2, \ldots, r-1)$ and the off diagonal blocks $B_{i+1,i}$ $(i = 1, 2, \ldots, r-2)$ are the only non-zero matrices;
   (3) $B_{i,i}$ $(i = 1, 2, \ldots, r-1)$ is a $(q-1) \times (q-1)$ matrix and

$$B_{i,i} = \begin{bmatrix} -1 & & & & \\ 1 & -1 & & & \\ & 1 & \ddots & & \\ & & \ddots & -1 & \\ & & & 1 & -1 \end{bmatrix};$$

   (4) $B_{i+1,i} = -B_{1,1}$ $(i = 1, 2, \ldots, r-2)$.

We can use the above matrix to calculate the Alexander polynomial of $K(q,r)$.

Theorem 7.3.2.
   Let $\Delta_{q,r}(t)$ be the Alexander polynomial of $K(q,r)$, $q, r \neq 0$.
   If g.c.d$(q,r) = d \geq 1$, i.e., $K(q,r)$ is a d-component torus link, then

$$t^{\frac{(q-1)(r-1)}{2}} \Delta_{q,r}(t) = (-1)^{d-1} \frac{(1-t)(1-t^{\frac{qr}{d}})^d}{(1-t^q)(1-t^r)}.$$

Since the calculation of $\det(M - tM^T)$ is a tad cumbersome, we shall not give a proof here but instead consider an example.

Example 7.3.1. We shall call the torus knot (or link) $K(n,2)$, the *elementary* torus knot (or link). Suppose $n > 0$ and let us calculate $\Delta_{K(n,2)}(t)$.

Let

$$t^{\frac{(n-1)}{2}}\Delta_{K(n,2)}(t) = \det \begin{bmatrix} -1+t & -t & & & \\ 1 & -1+t & -t & & \\ & 1 & \ddots & \ddots & \\ & & \ddots & -1+t & -t \\ & & & 1 & -1+t \end{bmatrix} = d_n.$$

A direct calculation shows that $d_2 = -1+t$ and $d_3 = 1-t+t^2$. We shall calculate d_{n+1} by induction on n. Suppose

$$d_n = (-1)^{n-1}(1-t+t^2-t^3+\cdots+(-1)^{n-1}t^{n-1}).$$

In order to determine d_{n+1}, first add all the rows to the first row, and then expand it along the first row so that we obtain

$$\begin{aligned} d_{n+1} &= td_n + (-1)^n \\ &= (-1)^{n-1}t(1-t+\cdots+(-1)^{n-1}t^{n-1}) + (-1)^n \cdot \\ &= (-1)^n(1-t+\cdots+(-1)^n t^n) \end{aligned}$$

Finally, it is easy to check that for $d = \gcd(n,2)$ $(= 1$ or $2)$,

$$d_n = (-1)^{d-1}\frac{(1-t)(1-t^{\frac{2n}{d}})^d}{(1-t^n)(1-t^2)}.$$

Exercise 7.3.1. Without using Theorem 7.3.2, calculate the Alexander polynomial of $K(3,4)$ and $K(3,3)$. Confirm your answers by using Theorem 7.3.2.

Exercise 7.3.2. Compute $\Delta_{q,r}$ and $\Delta_{-q,r}$ $(q,r \neq 0)$ using the formula given in Theorem 7.3.2 and confirm that $\Delta_{q,r} = \Delta_{-q,r}$.

Exercise 7.3.3. Show that there do not exist a q and r that will make the oriented torus link in Figure 7.1.8(c) equivalent to $K(q,r)$.

§4 The classification of torus knots (II)

As mentioned in the introduction to this chapter, to classify torus knots $K(q,r)$ $(q,r > 0)$ we already have the necessary techniques. In fact, all that is required is the Alexander polynomial.

Theorem 7.4.1.
    Suppose that $K(q,r)$ and $K(p,s)$ are two torus knots, and that $q, r, p, s \geq 2$. Then

$$K(q,r) \cong K(p,s) \Longleftrightarrow \{q,r\} = \{p,s\}.$$

Proof
    It follows from Theorem 7.2.4 that we may assume $q \geq r$ and $p \geq s$. Since the proof in the direction $\Longleftarrow$ is fairly immediate, we shall only show the proof in the direction $\Longrightarrow$, i.e., $q = p$ and $r = s$.
    We shall only prove the theorem for torus knots, so $q > r$ and $p > s$. The torus link case is left as an exercise for the reader.
    Now, since $K(q,r) \cong K(p,s)$, $\Delta_{K(q,r)}(t) = \Delta_{K(p,s)}(t)$. The maximum degree of $\Delta_{K(q,r)}(t)$ is

$$qr + 1 - (q+r) - \frac{(q-1)(r-1)}{2} = \frac{(q-1)(r-1)}{2},$$

while the maximum degree of $\Delta_{K(p,s)}(t)$, similarly, is

$$\frac{(p-1)(s-1)}{2}.$$

So, since $\Delta_{K(q,r)}(t) = \Delta_{K(p,s)}(t)$,

$$(q-1)(r-1) = (p-1)(s-1). \tag{7.4.1}$$

Therefore, due to Theorem 7.3.2, the following formula holds:

$$\frac{(1-t)(1-t^{qr})}{(1-t^q)(1-t^r)} = \frac{(1-t)(1-t^{ps})}{(1-t^p)(1-t^s)}. \tag{7.4.2}$$

If we clear the denominators in (7.4.2) and divide out by $(1-t)$, we obtain that

$$1 - t^p - t^s + t^{p+s} - t^{qr} + t^{p+qr} + t^{s+qr} - t^{p+s+qr}$$
$$= 1 - t^q - t^r + t^{r+q} - t^{ps} + t^{q+ps} + t^{r+ps} - t^{q+r+ps}. \tag{7.4.3}$$

    Let us, in both sides of (7.4.3) compare the minimum degree of the non-constant negative terms. Since $p > s$, the minimum degree of the negative terms on the left-hand side is either s or qr, while on the right-hand side the minimum degree is either r or ps. So it follows that there are four possible cases:

$$(i) \ s = r, \quad (ii) \ s = ps, \quad (iii) \ qr = r, \quad (iv) \ qr = ps.$$

However, since $p, q > 1$, the cases (ii) and (iii) cannot occur. So, suppose case (iv) holds, then since qr is the minimum degree we must have $qr \leq s < ps$. This now contradicts our assumption that $qr = ps$. Hence, we are left with only case (i), i.e., $s = r$. Then from (7.4.1) we also obtain the other required result, namely, $q = p$.

$\blacksquare$

In order to obtain a complete classification, we need to consider the cases when either q or r are negative. As a first step towards this, we shall prove that $K(q, r)$ is not amphicheiral.

Theorem 7.4.2.
The signature of $K(q, r)$ $(q, r > 1)$ is <u>not</u> zero. Therefore, $K(q, r)$ is not amphicheiral.

<u>Proof</u>
To actually determine the signature of $K(q, r)$ is quite a complicated procedure, cf. Theorem 7.5.1. However, to show that it is *not* zero is a slightly easier task. In this regard, let us look at the Seifert matrix M. We shall show that the signature of $P = -(M + M^T)$ is positive, i.e., the signature of $M + M^T$ is a negative integer. The proof is divided into two cases: r is an even integer, and r is an odd integer. We shall confine ourselves to showing the theorem is true in the cases of $K(4,4)$ and $K(4,5)$. The method to generalize and hence prove the theorem can easily be deduced from these two examples, and so we leave the generalization as an exercise for the reader.

(i) The case of the torus link $K(4,4)$.
For this case we can divide $P = -(M + M^T)$ into 9 blocks.

$$P = \begin{bmatrix} N + N^T & -N^T & O \\ -N & N + N^T & -N^T \\ O & -N & N + N^T \end{bmatrix},$$

where $N = \begin{bmatrix} 1 & 0 & 0 \\ -1 & 1 & 0 \\ 0 & -1 & 1 \end{bmatrix}$.

Each block is a matrix of order 3. For the sake of convenience, these blocks will be labeled the (i, j)-block of P [similar to the (i, j)-entry of a matrix] and denoted by $P_{i,j}$. For example, $P_{12} = -N^T$ and $P_{21} = -N$.

Next, we move $P_{2i+1,2j+1}$ $(i, j = 0, 1)$ to the top left-hand corner and $P_{2k,2l}$ $(k, l = 1)$ to the bottom right-hand corner. (This can be done by applying a Λ_1-operation on P.) The resultant matrix P' is therefore Λ_1-equivalent to P. To be more precise,

$$P' = \begin{bmatrix} P_{11} & P_{13} & P_{12} \\ P_{31} & P_{33} & P_{32} \\ P_{21} & P_{23} & P_{22} \end{bmatrix} = \begin{bmatrix} A & O & -N^T \\ O & A & -N \\ -N & -N^T & A \end{bmatrix},$$

where

$$A = N + N^T = \begin{bmatrix} 2 & -1 & 0 \\ -1 & 2 & -1 \\ 0 & -1 & 2 \end{bmatrix}.$$

Now, let $B = NA^{-1}$ and $C = N^T A^{-1}$. To clarify matters further, let us define a matrix, Q, of order 9,

$$Q = \begin{bmatrix} I & O & O \\ O & I & O \\ B & C & I \end{bmatrix},$$

where I denotes the third-order identity matrix, and use it to compute $QP'Q^T$,

$$QP'Q^T = \begin{bmatrix} A & O & O \\ O & A & O \\ O & O & A - BN^T - CN \end{bmatrix}.$$

Then $\sigma(P) = \sigma(QP'Q^T) = 2\sigma(A) + \sigma(A - BN^T - CN)$, where $\tilde{A} = A - BN^T - CN$ is a matrix of order 3. Hence $-3 \le \sigma(\tilde{A}) \le 3$, and since we may diagonalize A to

$$\begin{bmatrix} 2 & 0 & 0 \\ 0 & \frac{3}{2} & 0 \\ 0 & 0 & \frac{4}{3} \end{bmatrix},$$

it also follows that $\sigma(A) = 3$. Therefore, $\sigma(P) \ge 6 - 3 = 3 > 0$.

(ii) The case of the torus knot K(4,5).

If we apply a similar method to (i), we can only show that $\sigma(P) \ge 0$, so we need to slightly modify the proof of the previous case.

As in case (i), we may divide P into blocks, in this case into 16 blocks $P_{i,j}$ $(i,j = 1,2,3,4)$. Following along exactly the same lines as in case (i), we can move $P_{2i+1,2j+1}$ $(i,j = 0,1)$ to the top left-hand corner and $P_{2k,2l}$ $(k,l = 1,2)$ to the bottom right-hand corner. For the same reasons as before, the resultant matrix P' is Λ_1-equivalent to P:

$$P' = \begin{bmatrix} P_{11} & P_{13} & P_{12} & P_{14} \\ P_{31} & P_{33} & P_{32} & P_{34} \\ P_{21} & P_{23} & P_{22} & P_{24} \\ P_{41} & P_{43} & P_{42} & P_{44} \end{bmatrix} = \begin{bmatrix} A & O & -N^T & O \\ O & A & -N & -N^T \\ -N & -N^T & A & O \\ O & -N & O & A \end{bmatrix},$$

where A and N are as in (i).

As in (i), we may define a matrix Q using the same matrices B and C, but in this case the matrix is of order 12,

$$Q = \begin{bmatrix} I & O & & O \\ O & I & & \\ B & C & I & O \\ O & B & O & I \end{bmatrix}.$$

Also as in (i) we need to calculate $QP'Q^T$,

$$QP'Q^T = \begin{bmatrix} A & O & & O \\ O & A & & \\ & & A - BN^T - CN & -CN^T \\ O & & -BN & A - BN^T \end{bmatrix}.$$

Let us denote by R the right lower matrix (of order 6) of $QP'Q^T$. Then $\sigma(P) = \sigma(QP'Q^T) = 2\sigma(A) + \sigma(R)$.

Now, since

$$A^{-1} = \frac{1}{4} \begin{bmatrix} 3 & 2 & 1 \\ 2 & 4 & 2 \\ 1 & 2 & 3 \end{bmatrix},$$

a simple calculation shows that $BN^T = NA^{-1}N^T = N^TA^{-1}N = CN =$

$$\frac{1}{4} \begin{bmatrix} 3 & -1 & -1 \\ -1 & 3 & -1 \\ -1 & -1 & 3 \end{bmatrix}.$$

Since all the diagonal entries of R are positive, it follows from Exercise 6.4.1(a) that $-5 \leq \sigma(R) \leq 6$. Further, since $\sigma(A) = 3$, we have that $\sigma(P) = 6 + \sigma(R) \geq 6 - 5 = 1 > 0$. ∎

If we gather the above results together, then this leads to a classification of the torus knots.

Theorem 7.4.3 (Classification of torus knots).
(1) *If* q *or* r *is* 0 *or* ± 1 *then* $K(q,r)$ *is the trivial knot.*
(2) *Suppose that* q, r, p, s *are not* 0 *or* ± 1. *Then*

$$K(q,r) \cong K(p,s) \Longleftrightarrow \{q, r\} = \{p, s\} \text{ or } \{q, r\} = \{-p, -s\}.$$

Exercise 7.4.1. Generalize the above classification to include torus links $K(q,r)$. (Hint: Prove Proposition 7.2.1, Theorem 7.2.4, and Theorem 7.4.1 for torus links.)

§5 Invariants of torus knots

The proof of Theorem 7.4.2 seems to affirm that, in comparison with the Alexander polynomial, to actually calculate the signature of a torus knot (or link) is quite a complicated process. However, the following recurrence formula allows to calculate the signature in a more accessible way.

Theorem 7.5.1 [GLM].
Suppose that $q, r > 0$ *and denote* $-\sigma(K(q,r))$ *by* $\sigma(q,r)$, *then the following recurrence formula holds:*

(I) *Let us assume* $2r < q$, *then*
    (i) *if* r *is an odd integer,* $\sigma(q,r) = \sigma(q - 2r, r) + r^2 - 1$;
    (ii) *if* r *is an even integer,* $\sigma(q,r) = \sigma(q - 2r, r) + r^2$.

(II) $\sigma(2r,r) = r^2 - 1$.

(III) *Let us assume* $r \leq q < 2r$, *then*
    (i) *if* r *is an odd integer, then* $\sigma(q,r) + \sigma(2r - q, r) = r^2 - 1$;
    (ii) *if* r *is an even integer, then* $\sigma(q,r) + \sigma(2r - q, r) = r^2 - 2$;

(IV) $\sigma(q,r) = \sigma(r,q)$, $\quad \sigma(q,1) = 0$, $\quad \sigma(q,2) = q - 1$.

Example 7.5.1. Let us calculate $\sigma(14,5)$. First from (I)

$$\sigma(14,5) = \sigma(4,5) + 24 = \sigma(5,4) + 24.$$

Next, by (III)
$$\sigma(5,4) + \sigma(3,4) = 14.$$

However,

$$\sigma(4,3) + \sigma(2,3) = 8 \text{ and } \sigma(4,3) = \sigma(3,4) = 6.$$

Substituting these into the original equation gives $\sigma(14,5) = 32$.

Exercise 7.5.1. Calculate $\sigma(16,5)$ and $\sigma(16,6)$.

Exercise 7.5.2. Determine the general form for $\sigma(q,3)$. (Hint: Classify q into 6 residue classes.)

Exercise 7.5.3. Determine the signature of the oriented torus link of type (8,6) and compute $\xi(L)$ [for a definition of $\xi(L)$, see Exercise 6.4.7]. Using $\xi(L)$, determine the signature of the torus link L' obtained from L by reversing the orientation of only one component of L.

Exercise 7.5.4. Show that if q and r are both odd and $\gcd(q,r) = 1$, then $\sigma(q,r) \equiv 0 \pmod{8}$.

The signature of the torus knot of the type in Exercise 7.5.4 has been determined directly without recourse to the above recurrence formula in [HZ*]. In general, it is known that if $\Delta_K(-1) = 1$, then $\sigma(K) \equiv 0 \pmod{8}$.

We shall now look at several classical invariants for torus knots.

Theorem 7.5.2.
Let $\gcd(q,r) = 1$. The genus, g(K), of the torus knot $K(q,r)$ $(q,r > 0)$ is $\frac{(q-1)(r-1)}{2}$.

Proof
Suppose M is the Seifert matrix of $K(q,r)$ obtained from the Seifert surface constructed from the regular diagram in Figure 7.1.7. By Proposition 7.3.1,

$$\det M = (\det B_{11})(\det B_{22})\ldots(\det B_{r-1,r-1})$$
$$= [(-1)^{q-1}]^{r-1} \neq 0.$$

We know from Proposition 6.3.8 that g(K) is the maximum degree of the Alexander polynomial, which is equal to $\frac{(q-1)(r-1)}{2}$.

∎

As an aside, the Seifert surface constructed from the standard regular diagram, see Figure 7.1.7, is, in fact, a spanning surface of K with this (minimal) genus.

It follows easily from the regular diagram of $K(q,r)$ that $K(q,r)$ has a $|q|$-bridge presentation. However, since $K(q,r) \cong K(r,q)$ (cf. Theorem 7.2.4), $K(q,r)$ has, at worst, a $\min\{|q|,|r|\}$-bridge presentation. In fact, this is its bridge number.

Theorem 7.5.3 [Sc2].
  If q or r $\neq 0$, then

$$\mathrm{br}(K(q,r)) = \min\{|q|,|r|\}.$$

The regular diagram of $K(q,r)$ has exactly $|q|(|r|-1)$ crossing points. On the other hand, the regular diagram of $K(r,q)$ has exactly $|r|(|q|-1)$ crossing points. In fact, it has recently been proven that the minimum number of crossing points of $K(q,r)$ depends on these two numbers.

Theorem 7.5.4 [Mus6].
  The minimum number of crossing points of the torus link $K(q,r)$ is

$$c(K(q,r)) = \min\{|q|(|r|-1), |r|(|q|-1)\}.$$

Therefore, if $|q| \geq |r|$ the regular diagram of $K(q,r)$ in Figure 7.1.7 is the minimum regular diagram. To determine, on the other hand, the unknotting number of a torus knot is an extremely difficult problem

Exercise 7.5.5. Show that $u(K(q,2)) = \frac{|q|-1}{2}$. (Hint: Use Theorem 6.4.8.)

Exercise 7.5.6. For the torus knot $K(q,r)$ $(q,r > 0)$, show that

$$u(K(q,r)) \leq \frac{(q-1)(r-1)}{2}. \tag{7.5.1}$$

For a long time, it was believed that (7.5.1) could be improved if the stronger supposition "$\leq$" is replaced by equality. Quite recently, this supposition has been shown to be true.

Theorem 7.5.5 [KM].

For $q, r > 0$ and $\gcd(q, r) = 1$,

$$u(K(q, r)) = \frac{(q-1)(r-1)}{2}.$$

Exercise 7.5.7. Show that $K(q, r)$ has period $|q|$ and $|r|$. (For a definition of period, refer to Chapter 3, Section 2.)

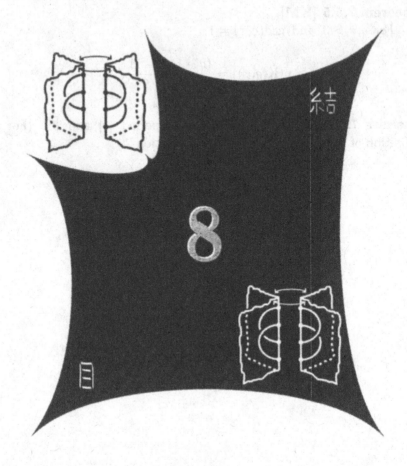

Creating Manifolds from Knots

So far in this book we have concerned ourselves with the problem of classifying knots (and, of course, links). Intrinsically, this is a knot theoretical problem. This book, however, is twofold in nature and we wish to balance the purely theoretical with some practical applications of knot theory. The various applications of knot theory are discussed in detail in the latter chapters of this book; we would, however, in this chapter like to consider what might be called the classic application of knot theory. One of the most important, even fundamental, problems in *algebraic topology* is the general classification of manifolds (see Definition 8.0.1 below). In this chapter we will show that it is possible to create from an arbitrary knot (or link) a 3-dimensional manifold (usually shortened to 3-manifold). Hence by studying the properties of knots we can gain insight into the properties of 3-manifolds.

A 3-manifold may be thought of as having one more dimension than a 2-manifold, which are more commonly known as *surfaces*. Surfaces, in general, are quite easy to visualize; however, once we add an extra dimension the actual shape can become ambiguous, and so visualizing a 3-manifold is not really possible, especially in the printed form. Hence, we shall in this chapter need to rely quite heavily on our imagination/geometrical awareness.

Definition 8.0.1. A 3-manifold is a topological space in which every point has a neighbourhood that is homeomorphic to the 3-ball.

For example, the 3-sphere S^3 and the ball B^3 are typical 3-manifolds that have already been encountered. To be precise, S^3 is a 3-manifold without boundary, while B^3 has a boundary that is homeomorphic to S^2. Also, the idea of a torus can be extended (exactly how we shall explain below) to a 3-dimensional torus – a 3-manifold without boundary.

One method to actually create a 3-manifold is to consider a solid polyhedron, i.e., like a ball its inside has been filled. The 3-manifold is made by suitably gluing to one another the polygons that make up the boundary of this polyhedron. This is essentially the analogue of the gluing process in 2 dimensions. The standard example in 2 dimensions is the torus, which is usually made by gluing together the two opposite sides of a square, see Figure 8.0.1.

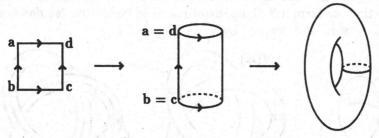

Figure 8.0.1

Carrying through this example into 3 dimensions, if we take a cube and in the "natural" way glue together the top and bottom faces, the front and back faces, and the left and right faces, then the resultant object is a 3-manifold called a 3-dimensional torus, see Figure 8.0.2.

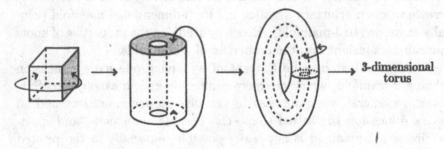

Figure 8.0.2

In this chapter, we shall explain two fundamental ways, both of which are closely related to knots, of constructing 3-manifolds. One approach is via Dehn surgery, while the other is via covering spaces. The fact that is of overwhelming importance concerning these two methods is that *every* closed connected orientable 3-manifold can be constructed by either of these methods. Therefore, in theory, we should be able to gain insight into the classification of 3-manifolds by using knot theory. However, at present the relationship is weighted in the opposite direction, more has been gleaned about knots by understanding their related 3-manifolds.

Throughout this chapter *all* the knots and links shall be considered to lie in S^3.

§1 Dehn surgery

Suppose that K is an (oriented) knot in S^3. Now, "slightly" thicken K, so that we form a 3-dimensional manifold called the *tubular neighbourhood* of K, and denoted by V(K), Figure 8.1.1.

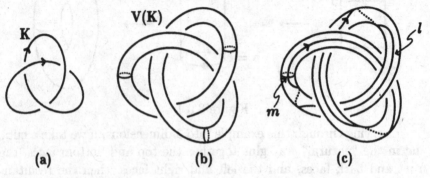

Figure 8.1.1

Since K is a knot, V(K) is a solid torus, and its boundary T [= ∂V(K)] is a torus. Therefore, we may define two characteristic simple closed curves, a meridian and longitude of T, as we discussed in Chapter 7, Section 2. However, due to the fact that we do not restrict ourselves solely to the trivial knot, T is not necessarily the trivial torus. Therefore, we have to slightly amend the definitions of our previous chapter. The definition of the meridian, m, can be naturally extended to the present case. The longitude, l, however, we shall fix as the intersection between the Seifert surface of K and T. [If the tubular neighbourhood V(K) is made sufficiently thin, then this intersection is a simple closed curve on T. Further, l is independent of the choice of the Seifert surface.] We may view l as being "parallel" to the knot K, which in this case is represented by the centre of V(K), Figure 8.1.1(c). We shall assign orientations to m and K in such a way that lk$(m, K) = 1$. The orientation of l is the same as the orientation of K, and, as noted above, we have assumed they are parallel.

Exercise 8.1.1. Show that the definition of the meridian m and the longitude l, given above, agree with our previous definition when T is the trivial torus. Further, show that lk$(l, K) = 0$.

By removing the interior points of V(K) from S^3, another 3-manifold $\widehat{V}$ is formed. $\widehat{V}$ and V(K) have a common boundary in T. For convenience, we shall denote the boundary of $\widehat{V}$ by $\widehat{T}$. The meridian and longitude that were defined on T may also be thought to lie on $\widehat{T}$ (and shall also be called a meridian and longitude of $\widehat{T}$).

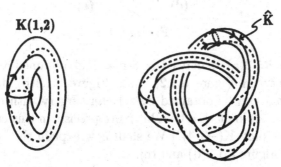

Figure 8.1.2

Now, let us take a trivial torus T_0 and let us place on T_0 a K(q, r)-torus knot. Let $h : T_0 \longrightarrow \widehat{T}$ $(= \partial\widehat{V})$ be a homeomorphism that sends the meridian and longitude of T_0 to the meridian and longitude of $\widehat{T}$. This homeomorphism also sends the torus knot on T_0 onto $\widehat{T}$. We may consider it to be a (q, r)-"torus" knot, $\widehat{K}$ on $\widehat{T}$. (In Figure 8.1.2, the case $q = 1$, $r = 2$ is shown.)

Next, let us consider $V(K)$ and $\widehat{V}$. Let $\hat{h} : \partial V(K) = T \longrightarrow \partial \widehat{V} = \widehat{T}$ be the auto-homeomorphism that sends the meridian, m, of $\partial V(K)$ to a torus knot, $\widehat{K}$ on $\partial \widehat{V}$. We now glue together, using this $\hat{h}$, the solid torus $V(K)$ to $\widehat{V}$ along their boundaries T and $\widehat{T}$. The resulting object is a (closed, orientable) 3-manifold M, which is said to have been constructed from S^3 by means of *Dehn surgery* along K, and $s = \frac{q}{r}$ is said to be its surgery coefficient. If instead of a knot we use a link $L = \{K_1, K_2, \ldots, K_\mu\}$, then again we can construct a 3-manifold, M. However, in this case M is obtained by performing Dehn surgery along each component. There exists a graphical notation for these 3-manifolds in which the surgery coefficient is placed next to the knot, Figure 8.1.3.

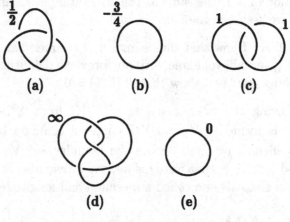

(a)   (b)   (c)

(d)   (e)

Figure 8.1.3

Figure 8.1.3(a) denotes the 3-manifold obtained by performing Dehn surgery on Figure 8.1.2, while Figure 8.1.3(b) denotes the 3-manifold that can be obtained by a Dehn surgery with coefficient $\frac{-3}{4}$ along the trivial knot. [To be a bit more precise, in this case m is sent to the torus knot $K(-3, 4)$.] We shall now explain the Dehn surgeries denoted in Figure 8.1.3(d) and (e).

Example 8.1.1. The Dehn surgery we perform has coefficient ∞. Since $\gcd(q, r) = 1$, if $r = 0$ the only possibility for q is ± 1. Then, we may fix $s = \frac{\pm 1}{0} = \infty$. Hence, in this case, the auto-homeomorphism $\hat{h}$ sends the meridian m of $\partial V(K)$ to the meridian m or $-m$ of $\partial \widehat{V}$. So this Dehn surgery is nothing but the operation that puts back the solid torus that we had previously removed. Therefore, the operation does not alter S^3, i.e., what is obtained after we have performed this Dehn surgery is the original S^3. Since by performing a Dehn surgery with coefficient ∞ along a knot (irrespective of the knot considered), we will always obtain S^3, in essence a trivial surgery, we shall not include such surgeries in our further discussions.

Example 8.1.2. Let us consider the 3-manifold M_0 denoted by Figure 8.1.3(e). Since the coefficient is zero, by definition we must have $r = \pm 1$, $q = 0$. Therefore, in this case the meridian m on $\partial V(K)$ is sent to the longitude $\pm l$ of $\partial \widehat{V}$. However, since K is the trivial knot, $V(K)$ is the trivial solid torus. By Proposition 7.2.2, $\widehat{V}$ is also a trivial solid torus. Furthermore, since the *meridian and longitude* of the solid torus $\widehat{V}$ become the *longitude and meridian*, respectively, of $\partial \widehat{V}$, we may think of M_0 as the 3-manifold obtained by gluing together, via the auto-homeomorphism $\widehat{h}$, the two trivial solid tori $\widehat{V}$ and $V(K)$ such that the meridian of $V(K)$ is sent to the meridian of $\widehat{V}$, Figure 8.1.4.

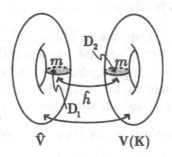

$\widehat{V}$ $V(K)$

Figure 8.1.4

These meridians are each the boundaries of disks D_1 and D_2 in $\widehat{V}$ and $V(K)$, respectively. If we glue these boundaries together, then the result is a sphere. Since this sphere lies along the longitude, M_0 is homeomorphic to $S^2 \times S^1$. Succinctly, Figure 8.1.3(e) denotes the 3-manifold $S^2 \times S^1$.

Figure 8.1.5

Example 8.1.3. The 3-manifold denoted by Figure 8.1.5 is called a (q,r)-Lens space and is denoted by $L(q,r)$. It is known that for any integers p and q, $L(p,q)$ and $L(-p,-q)$ are homeomorphic (with orientation preserved), and for any integer n, $L(p,q)$ and $L(p,q+np)$ are also homeomorphic (with orientation preserved). Therefore, we may assume that both p and q are non-negative integers. The study of Lens spaces has a long history, included in which is the complete classification of these spaces.

Theorem 8.1.1.

The necessary and sufficient conditions for two Lens spaces $L(q,r)$ *$(q,r>0)$ and $L(p,s)$ $(p,s>0)$, to be homeomorphic (with orientation preserved) are*

    (1) $q = p$

and

    (2) $r \equiv s \pmod{q}$ or $rs \equiv 1 \pmod{q}$.

For an interesting proof of this classical theorem using knot theory, we refer the reader to Brody [Br].

Example 8.1.4. The Lens spaces $L(3,4)$ and $L(3,1)$ are homeomorphic, but $L(7,3)$ and $L(7,4)$ are not homeomorphic (with orientation preserved). Moreover, if, in particular, k is an arbitrary integer, then $L(1,k)$ is homeomorphic to S^3, while $L(0,\pm1)$ is homeomorphic to $S^2 \times S^1$. (Note: k may also take the value zero.)

Finally in this section, let us state the fundamental theorem on Dehn surgery.

Theorem 8.1.2.

Any closed orientable connected 3-manifold can be obtained by performing a Dehn surgery to some knot (or link) in S^3. Further, we can take the surgery coefficient to be an integer.

Exercise 8.1.2. Show that Figure 8.1.3(c) denotes $S^2 \times S^1$.

From the above theorem it follows that every closed orientable connected 3-manifold corresponds to some link with an (integer) coefficient assigned to it. Since it is possible for two inequivalent links, with coefficients assigned, to represent the same 3-manifold [see, for example, Figures 8.1.3(c) and (e)], a classification of links will not induce a subsequent classification of 3-manifolds. However, in 1976 R. Kirby [Ki] introduced a concept of equivalence links with (Dehn surgery) integer coefficients. In that paper it was shown that there is a 1-1 correspondence between the equivalence type of links with integer coefficients and closed orientable connected 3-manifolds. Therefore, an invariant for links with coefficients assigned will pass through to become an invariant of 3-manifolds. In fact, there are numerous 3-manifold invariants that have been defined in this way.

§2 Covering spaces

In topology one of the most important concepts is a (topological) space called the *covering space*. Intuitively speaking, the covering space $\tilde{M}$ of a space M has locally the same structure as M. Globally, however, $\tilde{M}$ can be said to be a "large" space that covers M *equally* several times.

Example 8.2.1. Let us suppose M is the graph of the figure 8, Figure 8.2.1(a). *Locally*, this graph at its only vertex can be divided into four segments. The question now is, Which "large" graph possesses the very same characteristic? For example, Figure 8.2.1(b), (c), and (d) all have this characteristic but Figure 8.2.1(e) does not. Therefore (e) cannot be a covering space of (a). In fact, (b) ~ (d) are all covering spaces of (a).

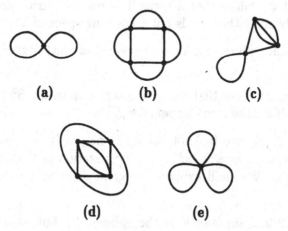

(a)   (b)   (c)

(d)   (e)

Figure 8.2.1

Let us now formally define a covering space.

Definition 8.2.1. Suppose X is an arcwise-connected topological space, i.e., any two points in X are joined by a simple arc in X. The arcwise-connected topological space $\widetilde{X}$ is a covering space of X if

(1) there exists a continuous map $p : \widetilde{X} \longrightarrow X$;
(2) for every point x of X there exists a (open) neighbourhood U_x of x such that
    (i) $p^{-1}(U_x)$ is the union of a family $F = \{W_\alpha\}$ of open sets of $\widetilde{X}$ that mutually do not intersect, i.e.,

$$(8.2.1)$$

$$p^{-1}(U_x) = \bigcup_{F \ni W_\alpha} W_\alpha.$$

    (ii) the restriction $p \,|\, W_\alpha$ for each W_α is a homeomorphism.

If the above conditions hold, then p is said to be a *covering map*. From (8.2.1) part (2)(*i*) the pre-image $p^{-1}(x)$ of a point x of X is a collection of isolated points; the number of points may be infinite. Also, the collection $p^{-1}(x)$ never includes segments. The *order* of this set of points is called the *degree* of the covering. Therefore, if $p^{-1}(x)$ has n points, then $\widetilde{X}$ is said to be the *n-fold covering space* of X.

Exercise 8.2.1. Confirm that Figures 8.2.1(b) $\sim$ (d) are covering spaces of Figure 8.2.1(a). Also, determine their degrees.

Exercise 8.2.2. Show that a torus T is not a covering space of S^2, and conversely show that S^2 is not a covering space of T.

Exercise 8.2.3. Show that a covering space of finite degree for a torus is a torus.

Exercise 8.2.4. Show that the only covering space for S^2 is of degree 1, i.e., itself. (Can the same be said for S^3 ?)

An important role in knot theory is played by a space called a *branched* covering space, which is a generalization of the concept of a covering space. We shall explain the generalization by means of an example.

Example 8.2.2. Suppose F is the sphere S^2, but with two small holes on the surface. Our intention is to make a 2-fold covering space for F. F is homeomorphic to a cylinder without a top and base (see Figure 8.2.2(a); the top and base correspond to the two holes α and β).

We can make a square $\widetilde{F}$ by cutting open F along a segment γ that connects two points A and B on F, where A and B are points on

α and β, respectively. Suppose $\widetilde{F}'$ is a copy of $\widetilde{F}$, Figure 8.2.2(b).
Now, glue together the right-hand edge γ of $\widetilde{F}$ to the left-hand edge
γ' of $\widetilde{F}'$, and the right-hand edge γ' of $\widetilde{F}'$ to the left-hand edge γ
of $\widetilde{F}$. The result of this process is a cylinder $M = \widetilde{F} \cup \widetilde{F}'$, which also
has no top and base, Figure 8.2.2(c). This *is* the 2-fold covering space
(covering surface) of F. To show this is in fact the case, we must find a
continuous map $p : M \longrightarrow F$ that satisfies (8.2.1). Firstly, from a point
P of F we can take the "same point" $\widetilde{P}$ of $\widetilde{F}$ and $\widetilde{P}'$ of $\widetilde{F}'$, which by
construction is just a copy of $\widetilde{F}$. Therefore, we can take p to be the
map that sends the two points $\widetilde{P}$ and $\widetilde{P}'$ on M to the point P on F.

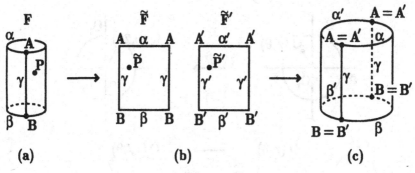

Figure 8.2.2

Exercise 8.2.5. Show that the above map $p : M \longrightarrow F$ satisfies (8.2.1).

If we close F by attaching two disks D_1 and D_2 to the top and
base, then we create a sphere $\widehat{F}$, Figure 8.2.3(a).

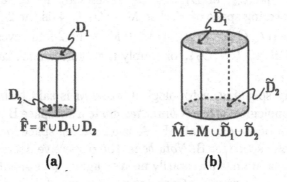

Figure 8.2.3

Similarly, we may create a sphere $\widehat{M}$ by attaching two disks $\widetilde{D}_1$

and $\widetilde{D}_2$ to the top and base of M, Figure 8.2.3(b). We may, as shown below, extend the previous continuous map p to a map $\hat{p}$ from this sphere $\widehat{M}$ to the sphere $\widehat{F}$. For the sake of clarity, suppose that $D_1, D_2, \widetilde{D}_1$, and $\widetilde{D}_2$ are all disks of the same size. Since $\widetilde{P}$ and $\widetilde{P}'$ are symmetric with regard to the axis of the cylinder, using polar co-ordinates we may define a map $\hat{p}_1 : \widetilde{D}_1 \longrightarrow D_1$, given by

$$\hat{p}_1(r, \theta) = (r, 2\theta),$$

see also Figure 8.2.4.

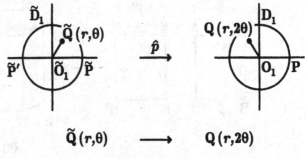

$$\widetilde{Q}(r, \theta) \longrightarrow Q(r, 2\theta)$$

Figure 8.2.4

Similarly, we can define a map $\hat{p}_2$ between the two base disks $\widetilde{D}_2$ and D_2. These two maps, $\hat{p}_1$ and $\hat{p}_2$, enable us to extend p to the continuous map $\hat{p} : \widehat{M} \longrightarrow \widehat{F}$. This map $\hat{p}$, excluding the centres of the disks $\widetilde{D}_1$ and $\widetilde{D}_2$, sends two points of $\widehat{M}$ to one point in $\widehat{F}$. The centres, O_1 and O_2, of D_1 and D_2 are in 1-1 correspondence with the centres, $\widetilde{O}_1$ and $\widetilde{O}_2$, of $\widetilde{D}_1$ and $\widetilde{D}_2$. Therefore, $\widehat{M}$ *to be exact* is not the 2-fold covering space of $\widehat{F}$, *but* $\widehat{M} - \{\widetilde{O}_1, \widetilde{O}_2\}$ is the 2-fold covering space of $\widehat{F} - \{O_1, O_2\}$. So we say that $\widehat{M}$ is the 2-fold covering space of $\widehat{F}$ *branched along* $\{O_1, O_2\}$, or simply the 2-fold *branch covering space* of $\widehat{F}$.

Roughly speaking, a topological space M is said to be a covering space of a topological space N *branched along* a subspace B if there exists a subspace A of M such that $M - A$ is a covering space for $N - B$, and A is a covering space for B. *Nota bene*, the respective degrees of the two covering spaces do not necessarily need to agree. (We should emphasize that this is only a rough interpretation, because if A is not connected then A may not be a covering space, an example of such a case is given in Exercise 8.4.1. For a precise definition we refer the reader to Fox [F1].)

§3 The cyclic covering space of a knot

The cyclic covering space, i.e., one of the covering spaces of S^3 branched along a knot, has been extensively studied over quite a number of years. Therefore, it has played a very important role in what can be said to be the first phase, i.e., the topological phase, of research into knot theory.

Suppose F is the Seifert surface of a knot K. Then K is the boundary of F. Let us now open up S^3 by cutting it open along F. The result is a 3-manifold, N, obtained from S^3, with a boundary. The boundary of N consists of two copies, F' and F'' of F, which have been glued together along K, Figure 8.3.1(a) and (b).

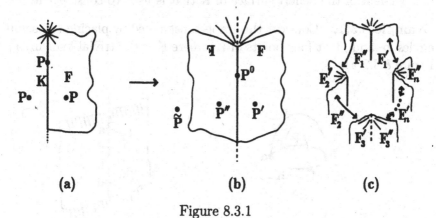

(a)   **(b)**   **(c)**

Figure 8.3.1

The correspondence between S^3 and N is as follows:
Let us suppose that P is a point of S^3.

(1) If P does not lie on F then P corresponds to the same point, $\widetilde{P}$, on N;

(2) If P is a point of K, then it corresponds to the same point P^o on the knot K on N;

(3) If P lies on F and is not a point of K, then it corresponds to exactly two points P' and P'' on the boundary of N. For the sake of clarity, let us denote the boundary of N with P' by F', and the boundary of N with P'' by F'', Figure 8.3.1(b). Now make n copies of N and denote them by $N_1, N_2, \ldots, N_n$. Let us suppose that P'_i and P''_i are copies of P' and P'', which lie on F'_i and F''_i, respectively. F'_i and F''_i are copies on N_i of F' and F''. We now glue together F''_1 (the boundary of N_1) and F'_2 (the boundary of N_2), so that the points P''_1 and P'_2 are in agreement. In a similar way, we glue together F''_2 and

F'_3, F''_3 and F'_4, ..., F''_{n-1} and F'_n, and finally F''_n and F'_1 in such a way that the points P''_i and P'_{i+1} $(i = 2, \ldots, n)$ are identified.

By carrying out this gluing and identifying process, what we should have constructed, finally, is a closed, connected, orientable 3-manifold, Figure 8.3.1(c). This 3-manifold is called the n-fold cyclic covering of S^3 branched along K and is denoted by $M_n(K)$. If K is the trivial knot, since F is a disk, N becomes the 3-ball. $M_n(K)$ is constructed by gluing the surfaces of these balls, i.e., it is homeomorphic to the 3-sphere S^3. The reason why we say it is "cyclic" is because the points

$$P'_1 \longrightarrow P''_1 = P'_2 \longrightarrow P''_2 = P'_3 \longrightarrow \cdots \longrightarrow P'_n \longrightarrow P''_n = P'_1$$

form a cycle. *Nota bene*, $M_n(K)$ depends only on the knot K and is independent of the Seifert surface of K that is used to construct it.

Example 8.3.1. Consider the graph constructed by placing two semicircles α and β at four points of K, where K is the trivial knot in S^3, Figure 8.3.2(a).

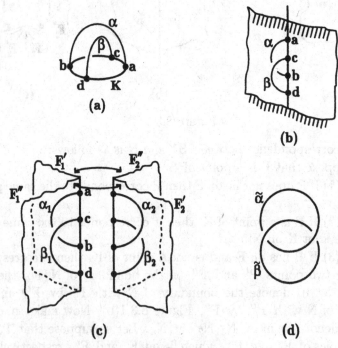

Figure 8.3.2

Suppose $M_2(K)$ is the 2-fold cyclic covering space of S^3 branched along K. It follows from our discussions above that $M_2(K)$ is also S^3, so let us see what happens to α and β in $M_2(K)$. Suppose the disk F is the Seifert surface of K. Then in the 3-ball N, obtained by cutting open S^3 along F, α and β are as shown in Figure 8.3.2(b). Therefore, the images α_1, β_1 and α_2, β_2 of α and β formed in the 3-balls N_1 and N_2, Figure 8.3.2, become, respectively, the closed curves $\tilde{\alpha} = \alpha_1 \cup \alpha_2$ and $\tilde{\beta} = \beta_1 \cup \beta_2$ in $M_2(K)$ ($\approx S^3$), which has been obtained by the above method of gluing F_1'' to F_2', and F_2'' to F_1'. So in N_2 they form a link, Figure 8.3.2(d).

Exercise 8.3.1. Using a similar method to Example 8.3.1, investigate what happens to the three arcs α, β, γ, shown in Figure 8.3.3, in the 2-fold cyclic covering of S^3 branched along K.

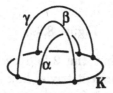

Figure 8.3.3

Exercise 8.3.2. (i) Show that the knot α in Figure 8.3.4 becomes the 3-component link $\tilde{\alpha}$, Figure 8.3.4(b), in $M_3(K)$ ($\approx S^3$), the 3-fold cyclic covering space of S^3 branched along K.

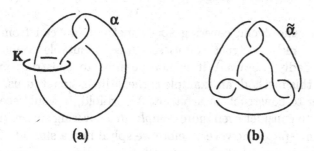

(a)   (b)

Figure 8.3.4

(ii) Similarly, show that the knot α in Figure 8.3.5(a) becomes the knot $\tilde{\alpha}$, Figure 8.3.5(b), in $M_3(K)$.

It is possible to deduce from Exercise 8.3.2 that the knot (or link) $\tilde{\alpha}$ created from a knot α that does not intersect the trivial knot K in the 3-fold cyclic covering space of S^3 branched along K has period 3. In general, it is possible to show that if $\widetilde{K}$ is a knot (or link) with period n (≥ 3), then $\widetilde{K}$ is equivalent to a knot (or link) in the n-fold cyclic covering space of S^3 branched along the trivial knot K_0 that has been transformed, by means of the above process from some other knot K_1 that does not intersect K_0. It is often the case that due to the above, the study of the period of a knot involves studying the links $\{K_0, K_1\}$ and the associated branched cyclic covering space.

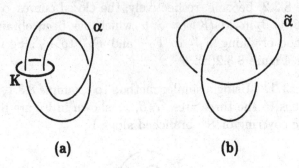

(a)   **(b)**

Figure 8.3.5

Exercise 8.3.3. Show that the 2-fold cyclic covering space of S^3 branched along the 2-component trivial link is $S^2 \times S^1$. [In general, it is known that the n-fold $(n \geq 2)$ cyclic covering space of S^3 branched along a μ-component $(\mu \geq 2)$ link is never homeomorphic to S^3.]

§4 A theorem of Alexander

By using the cyclic covering space method, we can from a single knot (or link) construct countless closed orientable connected 3-manifolds. Sadly, however, it is not possible to construct every 3-manifold by this method; an example is the 3-dimensional torus. Therefore, in order to construct an arbitrary 3-manifold, it would seem that we will need to consider even more *complicated* covering spaces than the cyclic covering space. However, before we spiral into a shroud of doom, actually we will find that the required covering space is surprisingly (relatively speaking) more simple than might be expected.

Theorem 8.4.1 (Alexander's theorem).

An arbitrary closed orientable connected 3-manifold can be constructed by means of a <u>3-fold</u> (in general non-cyclic) covering space of S^3 branched along some knot.

The proof of this theorem is given in Burde and Zieschang [BZ*], so here we shall explain how to construct such manifolds.

In the way described at the beginning of this chapter, we can construct a 3-manifold by gluing together the faces of a polyhedron. A covering space (of finite degree) can be constructed by generalizing the cyclic covering space method, i.e., we first make a (finite) number of copies of the solid polyhedron and then glue together, in a suitable manner, their faces. A requirement on the knot, used in Theorem 8.4.1 to construct a 3-fold branched covering space, is that its determinant is divisible by 3. In fact, we have already encountered such knots, namely, the knots that can be 3-coloured, cf. Chapter 4, Section 6. A typical example of such a knot is the trefoil knot; the figure 8 knot, however, is not suitable. Therefore, using the trefoil knot let us explain how to construct a covering space of the type in Theorem 8.4.1.

Firstly, let us take a point P that does not lie on K. We can now, in the obvious fashion by connecting (in S^3) P to each point of K, form a cone $\widehat{C}$, Figure 8.4.1(a).

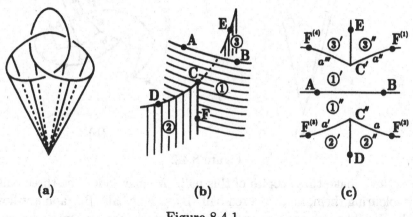

(a)   (b)   (c)

Figure 8.4.1

$\widehat{C}$ will intersect itself at the crossing points of the regular diagram of K. The intersection takes the form of a straight line, Figure 8.4.1(b). Next, open S^3 by cutting along $\widehat{C}$. Since $\widehat{C}$ intersects itself, at these points of intersection (straight lines) the subsequent shape is quite complicated; Figure 8.4.1(c) shows a section that has been opened in this way. As the result of cutting along the surface ①, the point C has now been divided into two points, C' and C''. Since the point F lies on the surfaces ① and ②, when we cut along the surfaces ① and ② it becomes divided into four points, $F^{(1)}$, $F^{(2)}$, $F^{(3)}$, and $F^{(4)}$. The points A, B, D, E, on the other hand, since they lie on the knot, will remain as single points after $\widehat{C}$ has been cut open. ①' and ①'' are the result of cutting along ①, i.e., the cutting process opens up ① into two parts.

Similarly, we obtain sections ②′, ②″, ③′, and ③″. Further, the segment that connects the vertex P of the cone to the point C divides into four parts. Suppose these four segments are a, a', a'', and a''', see Figure 8.4.1(c). Moreover, we may assign a front and back to the surfaces that have been cut open. (There is no specific rule to designate which is which. It is perfectly feasible to call the surface ①′ the front and the surface ①″ the back.) So we obtain from S^3 a ball that has on its surface a specific design. In Figure 8.4.2 we have shown the surface of the ball constructed for the case of the trefoil knot.

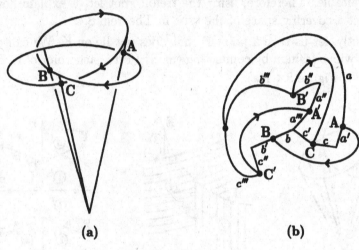

        (a)   **(b)**

Figure 8.4.2

Now, make three copies of this ball. We may index the three balls by colouring them, namely a red ball, B_r; a blue ball, B_b; and a yellow ball, B_y. The polygons on each of these surfaces are all pentagons, and they are glued together by means of the following rule. As we explained in Chapter 4, Section 6, we can colour K using 3 colours.

Suppose AB is part of a segment of the knot that is coloured red. Then,

(1) We may glue together, without alteration, the pentagons that are on either side of AB if these surfaces are part of the red ball B_r, Figure 8.4.3(a) (in other words, we restore the surface to its original form. If AB is blue (yellow), then do the same thing, but on this occasion the surfaces are part of the blue (yellow) ball B_b (B_y).

(2) If the pentagons that are on either side of AB are surfaces of the blue ball B_b and the yellow ball B_y, glue the pentagon that is on the front of B_b to the pentagon that is on the back of B_y in such a way that AB agrees [Figure 8.4.3(b)]. [If AB is blue (yellow), perform

a similar process, but now with B_r and B_y (B_r and B_b)].

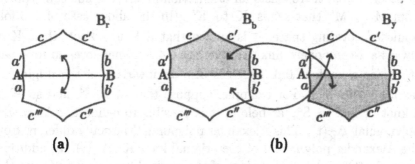

(a)   (b)

Figure 8.4.3

So, if we glue together the pentagons of all the surfaces of the three balls in the above prescribed manner, we shall obtain from B_r, B_b, and B_y a single 3-manifold. This 3-manifold is the 3-fold covering space of S^3 branched along K (with regard to the colouring on K). In fact, the 3-manifold constructed by this process from the trefoil knot is S^3.

Exercise 8.4.1. Show that the trefoil knot via the above process is transformed, in S^3, into a 2-component link.

Example 8.4.1. The 3-fold branched covering space obtained with regard to the 3-colouring of the knot in Figure 8.4.4 is a Lens space.

Figure 8.4.4

In Figure 8.4.4 the thick line denotes say the colour red, the wavy line the colour yellow, and the other line the colour blue.

The above method of construction need not be limited to 3-fold branched covering spaces, it is possible to generalize it to construct a covering space of finite degree branched along some knot.

If K and K' are equivalent knots, then M and M', the branched covering spaces (of the same type) constructed from K and K', respectively, are homeomorphic 3-manifolds. In M and M', the new knots (or links) $\widetilde{K}$ and $\widetilde{K}'$ that arise from K and K', respectively, are equivalent. In other words, there exists an orientation-preserving homeomorphism from M to M' that sends $\widetilde{K}$ to $\widetilde{K}'$. (In the above case of a 3-fold branched covering space, it is known that if K is a knot, then $\widetilde{K}$ is always a 2-component link, cf. Exercise 8.4.1.) Therefore, an invariant of $\widetilde{K}$ (as a knot or link in M) is also an invariant of the original K (as a knot in S^3). For example, suppose that M is S^3 and so $\widetilde{K}$ is a knot or link in S^3; as before, it is possible to define the Alexander polynomial $\Delta_{\widetilde{K}}(t)$. This Alexander polynomial is usually different from the Alexander polynomial of the original knot K, $\Delta_K(t)$. In addition, since $\Delta_{\widetilde{K}}(t)$ is also an invariant of the original knot K, we can use this second Alexander polynomial of K to help us classify knots.

In general, it is not easily possible to extend knot invariants in S^3; for example, the Alexander polynomial to knot invariants in an arbitrary 3-manifold M. However, for very simple invariants, for example, the linking number, it is possible to extend them to certain types of 3-manifolds.[13] These simple invariants are, in fact, powerful invariants of the original knot. These invariants can even, on occasion, distinguish two knots that have the same Alexander polynomial [BS].

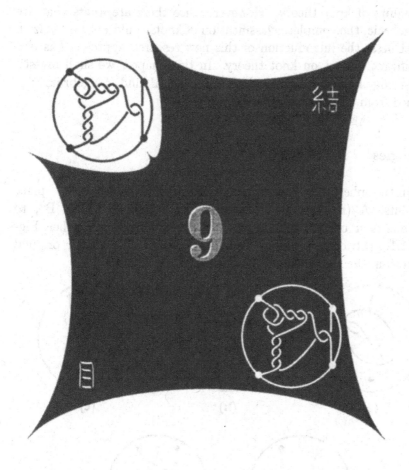

Tangles and 2-Bridge Knots

During the period from the end of the 1960s through to the beginning of the 1970s, Conway pursued the objective of forming a complete table of knots. As we have seen in our discussions thus far, the knot invariants that had been discovered up to that point in time were not sufficient to accomplish this aim. Therefore, Conway pulled another jewel from his bag of cornucopia and introduced the concept of a *tangle*. Using this variation on a knot, a new class of knots could be defined: algebraic knots. By studying this class of knots, various Local problems were able to be solved, which led to a further jump in the level of understanding of knot theory. However, since there are knots that are not algebraic, the complete classification of knots could not be realized. Nevertheless, the introduction of this new research approach has had a significant impact on knot theory. In this chapter we shall investigate 2-bridge knots (or links), which are a special kind of algebraic knot obtained from trivial tangles.

§1 Tangles

On the sphere S^2 – the surface (boundary) of the 3-ball B^3 – place $2n$ points. A (n, n)-tangle T is formed by attaching, within B^3, to these points n curves, none of which should intersect each other, Figure 9.1.1. (Strictly speaking, from the point of view of our original assumption, the curves should be polygonal.)

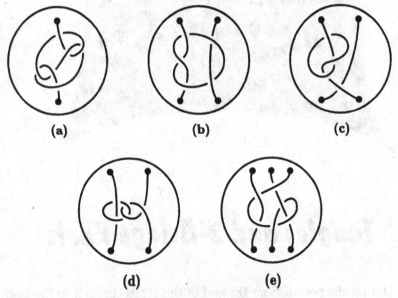

Figure 9.1.1

To be precise, we should say that a tangle is the set (B^3, T). However, since all our tangles will be within the ball B^3, we shall abbreviate the notation for a tangle to simply T. The astute reader may recall our discussions in Chapter 1, Section 5; the things discussed there should in keeping with the above definition be called $(1, 1)$-tangles. The case in which, in addition to our original n curves, there exists in B^3 a *closed* curve, Figure 9.1.1(d), shall not be considered in this book.

From the definition above, it follows immediately that Figure 9.1.1(a) is a $(1,1)$-tangle; Figure 9.1.1(b) $\sim$ (d) are $(2,2)$-tangles; and Figure 9.1.1(e) is a $(3,3)$-tangle. In what follows we will work virtually exclusively with $(2,2)$-tangles, and so for simplicity we shall refer to then as just *tangles*.

Suppose that we fix four points on the sphere S^2, namely, NE, NW, SE, SW (where the abbreviations are the obvious ones, north east, *et cetera*.), Figure 9.1.2.

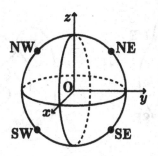

Figure 9.1.2

These points can be precisely described in $\mathbf{R}^3$ in terms of the following co-ordinates:

$$\mathrm{NE} = (0, \frac{1}{\sqrt{2}}, \frac{1}{\sqrt{2}}), \qquad \mathrm{NW} = (0, -\frac{1}{\sqrt{2}}, \frac{1}{\sqrt{2}}),$$

$$\mathrm{SE} = (0, \frac{1}{\sqrt{2}}, -\frac{1}{\sqrt{2}}), \qquad \mathrm{SW} = (0, -\frac{1}{\sqrt{2}}, -\frac{1}{\sqrt{2}}).$$

A cursory glance at these co-ordinates tells us that the four points all lie in the yz-plane. By attaching the end points of two polygonal curves in B^3 to these four points, we can form a tangle.

So, if we project this tangle onto the yz-plane, as in the case of a knot, we have what may be called a *regular* diagram of the tangle, Figure 9.1.3. (So as not to overcomplicate matters, we shall often consider a tangle to be simply this regular diagram.)

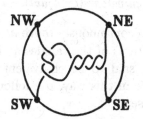

Figure 9.1.3

The knot (or link) obtained by connecting the points NW and NE, SW and SE by simple curves outside B^3, as is shown in Figure 9.1.4(a), is called the *numerator* and is denoted by N(T). Similarly, we may connect the points NW and SW, NE and SE by simple curves outside B^3, as is shown in Figure 9.1.4(b), and the subsequent knot (or link) is called the *denominator* and is denoted by D(T).

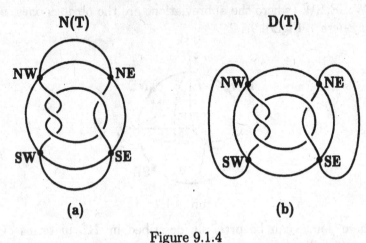

Figure 9.1.4

So, the above process allows us in a natural way to construct from a tangle two different knots (or links). The reader might be slightly confused with our new terms, since the numerator and denominator do not seem apt for the diagrams they represent. The reason for this somewhat strange terminology will be revealed slightly later (cf. Exercise 9.3.8).

Since a tangle is a "part" of a knot, we can extend the various definitions we have so far encountered to tangles, i.e., equivalence, the connected sum, *et cetera*.

Definition 9.1.1. Suppose T_1 and T_2 are two tangles in B^3. If we can change T_1 into T_2 by repeatedly performing elementary knot moves in B^3, keeping the four points (NE, NW, SE, SW) fixed, then T_1 and T_2 are said to be *equivalent* (or *equal*).

Intuitively, if we can continuously move in B^3 T_1 to T_2 without causing any self-intersections of the tangles and keeping the endpoints fixed, then T_1 may be said to be equivalent to T_2. Since we may think of knot equivalence in this way, the following definition is just a extension of the knot case:

Definition 9.1.2. If an orientation-preserving auto-homeomorphism

$\varphi : B^3 \longrightarrow B^3$ satisfies the following conditions, then T_1 and T_2 are said to be equivalent:

(1) φ is an identity map for S^2, i.e., the map keeps S^2 fixed.

(2) $\varphi(T_1) = T_2$.   (9.1.1)

It would seem obvious that there should be some way to connect two endpoints of one tangle to two endpoints of another tangle, in essence the *sum* of two tangles. To accomplish this, firstly place T_1 in a ball B_1^3 and T_2 in a ball B_2^3. Secondly, position T_1 and T_2 so that they become parallel, and form a "large" B_0^3 that contains of B_1^3 and B_2^3. Then we connect by parallel segments NE and SE of T_1 to NW and SW, respectively, of T_2. In this new configuration, the NW and SW of T_1 and NE and SE of T_2 become the NE, NW, SE, SW of the new ball B_0^3, and the "summed" tangle is denoted by $T_1 + T_2$, Figure 9.1.5.

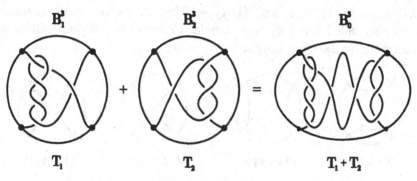

$$\mathbf{B_1^3} \qquad\qquad \mathbf{B_2^3} \qquad\qquad \mathbf{B_0^3}$$

$$\mathbf{T_1} \qquad + \qquad \mathbf{T_2} \qquad = \qquad \mathbf{T_1 + T_2}$$

Figure 9.1.5

We may think of this process in a slightly more precise fashion. Regard the ball as a globe, then after gluing the east hemisphere of B_1^3 to the west hemisphere of B_2^3, if we remove the parts of the hemispheres that have been glued together we shall form a ball. The "knotted" string in this ball is again the tangle sum.

It might seem at first sight that we may treat the sum of tangles in a similar way to the sum of knots (or links). However, there are considerably differences, as the next example and exercise show.

Example 9.1.1. Even if $N(T_1)$ and $N(T_2)$ are both non-trivial knots, it is quite possible that $N(T_1 + T_2)$ is a trivial knot.

This always occurs in the case of rational tangles, which we will define a bit later (see also Exercise 9.3.9).

Exercise 9.1.1. Show that the numerator of the tangles T_1 and T_2 in Figure 9.1.5 is not a trivial knot or link, but the numerator of $T_1 + T_2$ is a trivial knot.

When the surface of a 3-ball $\widehat{B}$ that lies in B^3 intersects the $(2,2)$-tangle T in only two points, then $(\widehat{B}, \widehat{B} \cap T)$ is a $(1,1)$-tangle. If this $(1,1)$-tangle is always the trivial tangle, then T is said to be *locally trivial* [for a definition of a trivial $(1,1)$-tangle, see Chapter 1, Section 5]. The tangles in Figure 9.1.1(b) and (d) are locally trivial tangles, while the one in (c) is not locally trivial.

§2 Trivial tangles (rational tangles)

Let us consider the simple tangles shown in Figure 9.2.1. Starting from the left-hand tangle, we shall call these tangles of (0)-type, $(0,0)$-type, (-1)-type, and (1)-type. The $(0,0)$-type tangle is sometimes also called the (∞)-type tangle. Collectively, we shall call these tangles the *exceptional* tangles.

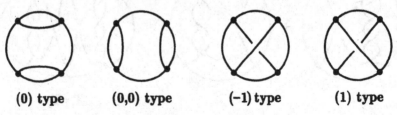

   (0) type   **(0,0) type**   **(−1) type**   **(1) type**

Figure 9.2.1

Exercise 9.2.1. Determine the knot (or links) that are the numerators and denominators of the exceptional tangles.

If we limit ourselves to $(1,1)$-tangles, it is obvious what should be the trivial $(1,1)$-tangle. However, it is not so straightforward to actually say without much thought what a trivial $(2,2)$-tangle is exactly. For example, we may say that the exceptional tangles are all "trivial" tangles. This, unfortunately, is not the limit of all possible trivial $(2,2)$-tangles; there are numerous other possibilities. Therefore, we need to give a formal definition of a *trivial tangle*.

Definition 9.2.1. Suppose f is a homeomorphism that maps the ball B^3 to itself and maps the *set* of four points $\{NW, NE, SW, SE\}$ to itself, but not necessarily as the identity map (i.e., f need not map NW to NW, *et cetera*). Then a *trivial tangle* (or *rational tangle*) is a tangle that is the image of the $(0,0)$-type tangle under this homeomorphism. (Hence, there are countless examples of such tangles.)

Example 9.2.1. If we rotate $\mathbf{R}^3$ about the x-axis through an angle of $\frac{\pi}{2}$, then the $(0,0)$-type tangle, $T(0,0)$, is sent to the (0)-type tangle, $T(0)$. Therefore, $T(0)$ is a trivial tangle, Figure 9.2.2.

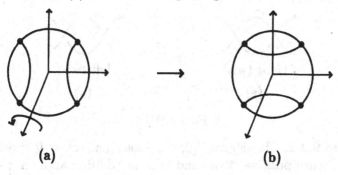

(a)   (b)

Figure 9.2.2

However, such "trivial" ones as the above are not the only trivial tangles. In fact, such homeomorphisms are numerous; as typical examples consider the following two examples.

Firstly, let us rotate the sphere about the z-axis, but keeping the northern hemisphere and the south pole fixed. Then the southern hemisphere is given a twist such that SE and SW exchange positions, Figure 9.2.3.

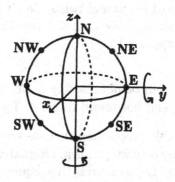

Figure 9.2.3

Secondly, let us consider the rotation of the sphere about the y-axis but on this occasion keeping the western hemisphere and the point $E(0,1,0)$ on the equator fixed. In this case, the eastern hemisphere is given a twist and the points NE and SE exchange positions, Figure 9.2.3. The former rotation/twist we shall call a *vertical twist*, and the latter a *horizontal twist*. Further, we may assign an orientation to the twists as described below. In the case of a vertical twist, a *positive twist* is a right twist [Figure 9.2.4(a)], while for a horizontal twist a *positive twist* is a left twist [Figure 9.2.4(b)]. The respective inverse twists are the negative twists.

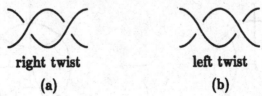

right twist   **left twist**

(a)   **(b)**

Figure 9.2.4

Example 9.2.2. If we give T(0,0) a 3-fold (i.e., rotate it three times) vertical (hence positive) twist and then in addition apply a (-4)-fold horizontal twist, then we obtain the tangle in Figure 9.1.3. Therefore, this is also a trivial tangle. In particular, T(1) and T(-1) are trivial tangles.

Example 9.2.3. We can obtain the tangle T_1 of Figure 9.1.5 by first sending T(0,0) to T(0) and then performing 2 horizontal twists, 3 vertical twists, and finally 1 horizontal twist in that order. Therefore, T_1 is also a trivial tangle.

The trivial tangles that we have obtained depend on performing alternatively vertical and horizontal twists, i.e., it is possible to formulate the following proposition. (We omit the proof since it depends on the theory of surface mappings.)

Proposition 9.2.1.
A trivial tangle can be obtained by performing a finite alternative sequence of vertical and horizontal twists to T(0) or T(0,0).

It follows from this proposition that the trivial tangles can be completely determined by how we perform, alternatively, the several vertical and horizontal twists. Let us express this sequence by T$(a_1, a_2, \ldots, a_n)$ and explain in a little more detail how this sequence occurs.

If n is odd, we can obtain the relevant tangle by first performing a_1 horizontal twists on T(0), then a_2 vertical twists, a_3 horizontal twists, and, continuing in this vein, repeating alternatively the twists until finally we perform a horizontal twist a_n times; an example is given in Figure 9.2.5(b).

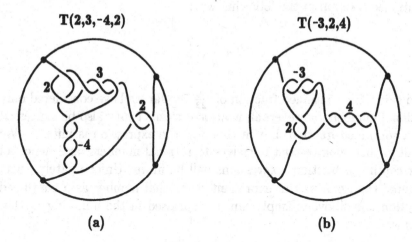

T(2,3,-4,2)   **T(-3,2,4)**

(a)   (b)

Figure 9.2.5

If n is even, we can obtain the relevant tangle by first performing a_1 vertical twists on $T(0,0)$, then a_2 horizontal twists, and then alternating the twists until finally we perform a horizontal twist a_n times; see also Figure 9.2.5(a).

If all the a_i are of the same sign, then the regular diagram that is obtained is an alternating diagram. The regular diagram is no longer alternating if the signs of some a_i and a_{i+1} are different. We also allow the case of a_i equal to zero. However, since if a_i is zero we can "shorten" the tangle, we shall, to avoid unnecessary complications, assume that $a_i \neq 0$ $(i \neq n)$.

Exercise 9.2.2. Show that $T(2,-3,0,2,1,-2)$ and $T(2,-1,1,-2)$ are equivalent.

A nice piece of number theory states that a real number may be expressed as a continued fraction. An example of such a continued fraction is

$$2 + \cfrac{3}{4 + \frac{2}{5}}.$$

From the "fraction" we can calculate the real number it expresses, starting from the last fraction in the sequence,

$$2 + \cfrac{3}{4 + \frac{2}{5}} = 2 + \cfrac{3}{\frac{22}{5}} = 2 + \cfrac{15}{22} = \cfrac{59}{22}.$$

Hence, the above continued fraction is equal to $\frac{59}{22}$.

However, it will help us in our further discussions if we express the continued fraction in the following way:

$$2 + \frac{3}{4} + \frac{2}{5}$$

and call it the continued fraction of $\frac{59}{22}$. So far we have considered only rational numbers; however, an irrational number may also be expressed as a continued fraction, but in this case the expression is infinite. We shall in this book restrict ourselves to rational numbers, and hence all our continued fraction, expressions will be finite. Unfortunately, there is more than one way of expressing a rational number as a continued fraction, our above example can be expressed in the following further two ways.

$$\frac{59}{22} = 2 + \frac{1}{1} + \frac{1}{2} + \frac{1}{7} = 3 + \frac{1}{(-3)} + \frac{1}{(-7)}.$$

In particular, if except the initial integer, the numerator of all the fractions is 1; then we shall express it as $[2, 1, 2, 7]$ $(= [3, -3, -7])$.

Exercise 9.2.3. Find at least 3 continued fraction expansions for $\frac{251}{32}$ and $\frac{11}{3}$.

The trivial tangle $T(a_1, a_2, \ldots, a_n)$, where $a_1 \neq 0$, corresponds to the fraction $\frac{p}{q}$ that has a continued fraction expansion given by

$$[a_n, a_{n-1}, \ldots, a_2, a_1] = \frac{p}{q}.$$

This number is called the *fraction of the tangle*. In particular, the fraction of the tangle $T(0)$ is 0, and the fraction of the tangle $T(0,0)$ is $0 + \frac{1}{0}$, which we will denote by ∞. Conversely, given a rational number p/q by expressing it in terms of a continued fraction,

$$[a_n, a_{n-1}, \ldots, a_2, a_1] \quad (a_i \neq 0),$$

we can associate it with the trivial tangle $T(a_1, a_2, \ldots, a_n)$. (Care needs to be taken with the order of $a_1, \ldots, a_n$.) The next theorem shows that there exists a very nice correspondence between the rational numbers (including ∞) and the trivial tangles.

Theorem 9.2.2 [C].

There exists a 1-1 correspondence between the set of all rational numbers (including ∞) and the equivalence classes of the trivial tangles. In other words, if the trivial tangles $T(a_1, a_2, \ldots, a_n)$
and $T(b_1, b_2, \ldots, b_n)$ are equivalent, then their respective fractions
that are expressed by the corresponding continued fraction expressions
$[a_n, a_{n-1}, \ldots, a_2, a_1]$ and $[b_m, b_{m-1}, \ldots, b_2, b_1]$ are equal. The converse also holds.

Since an arbitrary rational number (including ∞) corresponds to
a trivial tangle, the trivial tangles are sometimes referred to as *rational
tangles*.

Exercise 9.2.4. Show the validity of Theorem 9.2.2 by showing
$T(2,1,1,0)$ and $T(-2, -2,1)$ are equivalent; this should follow since
$3/5$ has the the continued fraction expansions $[0,1,1,2]$ and $[1, -2, -2]$.

If in the process of determining the rational number from a given
continued fraction, we set $\frac{1}{0} = \infty$, $\frac{1}{\infty} = 0$, and $k + \infty = \infty$, then this
will obviate any problems that might occur if any of the a_i are zero.

Exercise 9.2.5. Show using regular diagrams that $T(0, 3, 0) = T(0)$
and $T(1, -1, 2) = T(0, 0)$.

Exercise 9.2.6. Show that if the trivial tangle $T(a_1, a_2, \ldots, a_n)$ has
as its fraction a rational number other than 0 or ∞, then we can always
find a_i such that the signs of all the a_i are the same.

The sum of two rational tangles is not necessarily a rational tangle.
Figure 9.2.6(c) is the sum of two rational tangles but itself is not a
rational tangle.

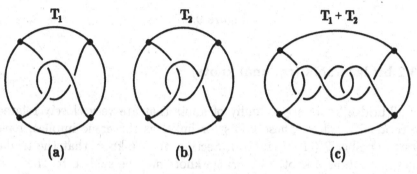

Figure 9.2.6

In fact, from Theorem 9.3.1, which we will prove in the next section, the denominator of a rational tangle is always a prime knot (or link), but the denominator of $T_1 + T_2$ of Figure 9.2.6(c) is not prime. We say a tangle is an *algebraic tangle* if it can be expressed as the sum of a finite number of tangles, each of which is a rational tangle and/or its homeomorphic image (mirror image, rotation, *et cetera*). Although we will give a few comments concerning algebraic tangles at the end of this chapter, we shall not delve too deeply into the concept of algebraic tangles. A simple example of an algebraic tangle is given in the final example of this section.

Example 9.2.4. From Figure 9.2.7, since T_1' is T_1 rotated by $\frac{\pi}{2}$, the sum of T_1' and T_2, $T_1' + T_2$, is an algebraic tangle.

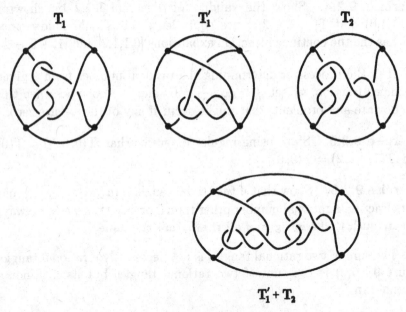

$$T_1' + T_2$$

Figure 9.2.7

§3 2-bridge knots (rational knots)

2-bridge knots are a family of knots that are very closely related to rational tangles. These knots (or links) as the name implies, have bridge number 2 (cf. Chapter 4, Section 3). We know that the trivial knot is a 1-bridge knot, so 2-bridge knots may be said to be the next most simplest set of knots (and links) to investigate. 2-bridge knots (or links) are always prime and have at most 2 components (why?). Further, this set of knots and links has been completely classified, but the local characteristics of these knots (or links) are still an important area of research. Since a non-trivial (q, r)-torus knot has bridge number $\min\{|q|, |r|\}$, if q or r is not ± 2, they do not belong to this class of knots.

Our first task in this section is to show the relation between 2-bridge knots and rational tangles. To this end we shall prove the next theorem. *Caveat lector*, in this section we will only consider knots and links that are *not* oriented.

Theorem 9.3.1.

(1) *A 2-bridge knot (or link) is the denominator of some rational tangle.*

(2) *Conversely, the denominator of a rational tangle is a 2-bridge knot (or link).*

Due to the above theorem, 2-bridge knots are often called *rational knots.*

<u>Proof</u>

Let us consider a regular diagram, D, of a 2-bridge knot (or link). Since D has only two local maxima, the regular diagram may be drawn as the reduced regular diagram shown in Figure 9.3.1(a); refer to Theorem 4.3.3. (In the diagram, the local maxima occur, sideways, at the left-hand edge.)

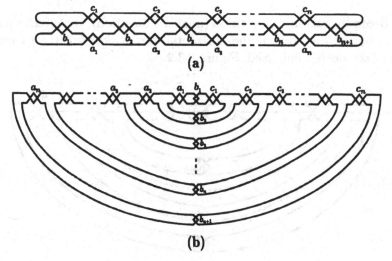

(a)

(b)

Figure 9.3.1

Our aim is to move the crossing points on the bottom so that in the diagram all the crossing points are in the centre or on top. (Since the proof does not rely on the crossing information at the crossing points, we shall not distinguish between over- and under-crossings. So, let us denote the number of crossing points by a_i, b_j, c_k. We allow the possibility that some of a_i, b_j, c_k are zero.)

Now, we deform the diagram of Figure 9.3.1(a) into the diagram of Figure 9.3.1(b). For the sake of convenience, the diagram D in Figure 9.3.1(b) will be denoted by the following notation:

$$D = (a_1, a_2, \ldots, a_n \mid b_1, b_2, \ldots, b_n, b_{n+1} \mid c_1, c_2, \ldots, c_n).$$

Explicitly, we want to move all the $a_1, a_2, \ldots, a_n$ crossing points on the left to the right. In other words, we want to show that D can be deformed into a new diagram

$$(0, 0, \ldots, 0 \mid b_1, b_2, \ldots, b_n, b_{n+1} \mid a_1 + c_1, a_2 + c_2, \ldots, a_n + c_n).$$

For simplicity, let $d_i = a_i + c_i$ $(i = 1, 2, \ldots, n)$, and call the half-circular band with b_i crossing points the ith *arm* of D.

First, move the a_n, the left-most crossing point, to the right by twisting the $(n+1)^{th}$ arm either right or left an appropriate number of times.

Thus, D is deformed into

$$D_1 = (a_1, a_2, \ldots, a_{n-1}, 0 \mid b_1, b_2, \ldots, b_n, b_{n+1} \mid c_1, c_2, \ldots, d_n).$$

Secondly, rotate, a_{n-1} times (in either direction), the interior part of the dotted line around the horizontal axis A, keeping the exterior part of the dotted line fixed, Figure 9.3.2.

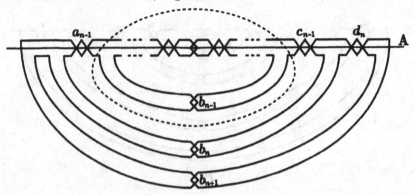

Figure 9.3.2

Thus, a_{n-1} crossing points have been moved to the right. Such a rotation will hereafter be called a *horizontal rotation*.

There are two cases to consider.

(I) If a_{n-1} is even, then the resulting diagram is

$$D_2 = (a_1, a_2, \ldots, a_{n-2}, 0, 0 \mid b_1, \ldots, b_n, b_{n+1} \mid c_1, c_2, \ldots, d_{n-1}, d_n).$$

(II) If a_{n-1} is odd, then

$$D_2 = (a_1, a_2, \ldots, a_{n-2}, 0, 0 \mid b_1, \overline{b_2}, \ldots,$$
$$\overline{b_{n-1}}, b_n, b_{n+1} \mid c_1, c_2 \ldots, c_{n-2}, d_{n-1}, d_n),$$

where $\overline{b_i}$ means that the ith arm with b_i crossing points is above the horizontal axis, as shown in Figure 9.3.3.

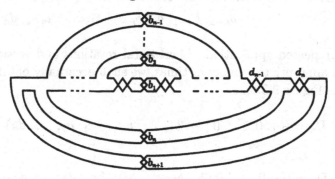

Figure 9.3.3

Now to lower the $(n-1)^{\text{th}}$ arm, we shall form a new horizontal $(n-1)^{\text{th}}$ arm (without crossing points) under A, Figure 9.3.4.

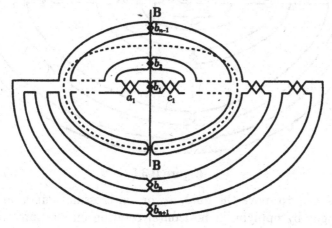

Figure 9.3.4

Next, rotate, b_{n-1} *times* the interior of the dotted line, in Figure 9.3.4, around the vertical axis B so that the b_{n-1} crossing points that lie above A disappear, but b_{n-1} crossing points are created in the lower arm. Such a rotation will be called a vertical rotation.

The resulting diagram, D_3 is of the form:

(a) if b_{n-1} is even, then

$$D_3 = (a_1, a_2, \ldots, a_{n-2}, 0, 0 \mid b_1, \overline{b_2}, \ldots,$$
$$\overline{b_{n-2}}, b_{n-1}, b_n, b_{n+1} \mid c_1, \ldots, c_{n-2}, d_{n-1}, d_n).$$

(b) if b_{n-1} is odd, then

$$D_3 = (c_1, c_2, \ldots, c_{n-2}, 0, 0 \mid b_1, \overline{b_2}, \ldots,$$
$$\overline{b_{n-2}}, b_{n-1}, b_n, b_{n+1} \mid a_1, \ldots, a_{n-2}, d_{n-1}, d_n).$$

By repeated application of horizontal rotations and vertical rotations an appropriate number of times, we shall eventually obtain a new diagram, Figure 9.3.5,

$$D_n = (a_1, 0, \ldots, 0 \mid b_1, b_2, \ldots, b_{n+1} \mid c_1, d_2, \ldots, d_n)$$

or

$$D_n = (c_1, 0, \ldots, 0 \mid b_1, b_2, \ldots, b_{n+1} \mid a_1, d_2, \ldots, d_n).$$

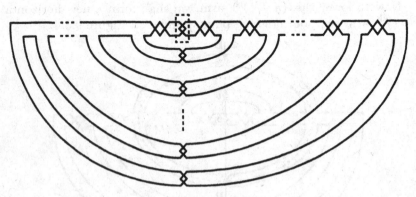

Figure 9.3.5

It is easy to move the final a_1 or c_1 crossing points on the left to the right by applying a horizontal rotation on the interior of the dotted square.

Exercise 9.3.1. Confirm that the procedures outlined in the above part of the proof work by performing them on the regular diagram in Figure 9.3.6.

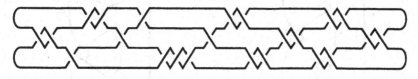

Figure 9.3.6

Therefore, we may think of a *standard* regular diagram for a 2-bridge knot (or link) as the one shown in Figure 9.3.7.

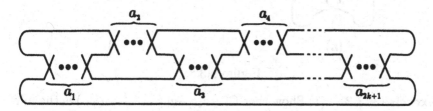

Figure 9.3.7

The number of twists is denoted by the integer a_i, and we can define the sign of a_i as follows:

If i is odd then the right twist is positive, if i is even then the left twist is positive (cf. Figure 9.2.4).

The regular diagram (Figure 9.3.7) may also be considered to be the denominator of the rational tangle $T(a_1, a_2, \ldots, a_{2k+1}, 0)$. So Theorem 9.3.1(1) has been proven. Conversely, it can be easily seen that the denominator of a rational tangle has a regular diagram that is the standard regular diagram of a 2-bridge knot (or link), such as for example, in Figure 9.3.7.

∎

We say that a 2-bridge knot (or link) that has a standard regular diagram of the form in Figure 9.3.7 is a 2-bridge knot (or link) of $(a_1, \ldots, a_{2k+1})$ type, and we shall denote it by $C(a_1, a_2, \ldots, a_{2k+1})$.

Exercise 9.3.2. Show that the 2-bridge knot $C(4, -2, 5, 1, -2)$ is $D(T(4, -2, 5, 1, -2, b))$, where b is an arbitrary integer. Further, by changing the regular diagram, show that it is equivalent to $C(3,2,6)$, see Figure 9.3.8.

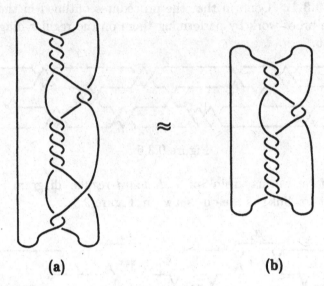

(a)   **(b)**

Figure 9.3.8

Exercise 9.3.3. (1) Show that $C(-a_1, -a_2, \ldots, -a_{2k+1})$ is the mirror image of $C(a_1, a_2, \ldots, a_{2k+1})$.
          (2) Show that

$$C(a_1, a_2, \ldots, a_{2k+1}) \approx C(a_{2k+1}, \ldots, a_2, a_1).$$

We may assume, due to Exercise 9.2.6, that the signs of all the a_i of a rational tangle, $T(a_1, a_2, \ldots, a_{2k+1})$, excluding exceptional tangles, are the same. Therefore, this regular diagram is an alternating diagram.

Proposition 9.3.2.
   2-bridge knots (or links) are alternating.

Exercise 9.3.4. Prove Proposition 9.3.2 without using Theorem 9.3.1 and Exercise 9.2.6. What needs to be shown is that the standard diagram of a 2-bridge knot (or link) can be deformed into an alternating diagram. In particular, find an alternating diagram of the knot in Figure 9.3.6.

In what follows, we shall consider only 2-bridge knots (or links) that are obtained from non-exceptional rational tangles.

In a similar way to a rational tangle, we can associate a 2-bridge knot (or link) $C(a_1, a_2, \ldots, a_{2k+1})$, with $a_i \neq 0$, with a rational number

$$\frac{\alpha}{\beta} = [a_1, a_2, \ldots, a_{2k+1}], \quad \gcd(\alpha, \beta) = 1.$$

(*Nota bene*, care must be taken when dealing with the continued fraction expansion.)

If all the a_i are positive integers, then $0 < \beta < \alpha$. However, all the a_i are negative, since $\frac{\alpha}{\beta} < -1$, we may assume that $\alpha > 0$ and $\beta < 0$. We then say that (α, β) is the type of $C(a_1, a_2, \ldots, a_{2k+1})$. Hence, any 2-bridge knot (or link) corresponds to some (α, β), where $-\alpha < \beta < \alpha$ and $\alpha > 0$.

Example 9.3.1. The type of the 2-bridge knots in Figures 9.3.8(a) and (b) is (45,13). In this case, (a) and (b) correspond to the continued fractional expansions $[4, -2,5,1, -2]$ and $[3,2,6]$, respectively, of $\frac{45}{13}$.

Conversely, from a rational number $\frac{\alpha}{\beta}$ $(|\frac{\alpha}{\beta}| > 1)$, we can create a 2-bridge knot (or link) in the way described below.

Let us assume that $0 < \beta < \alpha$. Firstly, let us look for a continued fraction expansion $[a_1, a_2, \ldots, a_m]$ for $\frac{\alpha}{\beta}$ such that $a_i > 0$.

If m is even and $a_m > 1$, then we write

$$[a_1, a_2, \ldots, a_m] = [a_1, a_2, \ldots, a_m - 1, 1]$$

In the case $a_m = 1$, we may rewrite it as

$$[a_1, a_2, \ldots, a_m] = [a_1, a_2, \ldots, a_{m-1} + 1]$$

From the above continued fraction expansion of $\frac{\alpha}{\beta}$, we may always assume that m is odd. If, in fact, we assume this, then we shall obtain a correspondence between the 2-bridge knot $C(a_1, a_2, \ldots, a_m)$ and $\frac{\alpha}{\beta}$. Therefore, the type of this 2-bridge knot (or link) is (α, β). (In the case when $\beta < 0$, the subsequent correspondence for the 2-bridge knot has all the a_i negative). If, from two different continued fraction expansions of $\frac{\alpha}{\beta}$, we obtain regular diagrams of two 2-bridge knots, then they are regular diagrams for the same knot. In short, 2-bridge knots (or links) are completely classified by their type (α, β), as the following theorem clarifies:

Theorem 9.3.3.
Suppose that K *and* K' *are 2-bridge knots (or links) of type* (α, β) *and* (α', β'), *respectively. Then* K *and* K' *are equivalent (excluding orientation in our considerations) if and only if the following holds:*
(1) $\alpha = \alpha'$, $\beta \equiv \beta' \pmod{\alpha}$
or
(2) $\alpha = \alpha'$, $\beta\beta' \equiv 1 \pmod{\alpha}$

Further, the mirror image K^* of K is a 2-bridge knot of type $(\alpha, -\beta)$. Therefore, a necessary and sufficient condition for K to be amphicheiral is that

(3) $\beta^2 \equiv -1 \pmod{\alpha}$.

The essence of the proof of the above theorem relies on the fact that the 2-fold cyclic covering space branched along K is a Lens space of type (α, β). Then Theorem 8.1.1 implies Theorem 9.3.3. Besides this observation, we will not plow into the details of the proof, but direct the reader to Schubert [Sc3] for a more detailed discussion.

It is also known but again we will spare the reader at this juncture the exact details of another proof that the determinant of a 2-bridge knot (or link) K is α. Therefore, 2-bridge knots cannot solely be classified by means of their determinants. (Note: The determinant of K does not depend on the orientation.)

Exercise 9.3.5. Show by determining their types that the two knots $C(1,1,1,2,1)$ and $C(2,4,1)$ are not equal. (These two knots have the same knot determinant.)

Exercise 9.3.6. (1) Find the 2-bridge knots that correspond to $\frac{23}{6}$ and $\frac{23}{5}$, and determine whether they are equivalent.

(2) As in (1) but with regard to $\frac{29}{13}$ and $\frac{29}{9}$.

Exercise 9.3.7. Let K be a 2-bridge knot (or link) of type (α, β). Show that K is a knot if and only if α is odd.

So far we have shown that the denominator of a rational tangle $T(a_1, a_2, \ldots, a_{2k+1})$ is a 2-bridge knot (or link). The natural question is, What can we say about the numerator? The next proposition shows, by comparing various regular diagrams, that the numerator is also a 2-bridge knot (or link).

Proposition 9.3.4.

For any integer b, the following hold:

$$(1) \ N(T(a_1, a_2, \ldots, a_{2k+1})) \approx N(T(a_1, a_2, \ldots, a_{2k+1}, b, 0))$$
$$\approx D(T(-a_1, -a_2, \ldots, -a_{2k+1}, b))$$
$$\approx C(-a_1, -a_2, \ldots, -a_{2k+1}).$$

$$(2) \ N(T(a_1, a_2, \ldots, a_{2k})) \approx D(T(-a_1, -a_2, \ldots, -a_{2k}, b))$$
$$\approx C(a_1, a_2, \ldots, a_{2k} - 1, 1).$$

(3) $D(T(a_1, a_2, \ \ldots \ , a_{2k})) \approx D(T(a_1, a_2, \ldots, a_{2k-1}, 0))$

$\approx C(a_1, a_2, \ldots, a_{2k-1}).$

(4) $D(T(a_1, a_2, \ldots, a_{2k+1})) \approx D(T(a_1, a_2, \ldots, a_{2k}, 0))$

$\approx C(1, a_1 - 1, a_2, \ldots, a_{2k}).$

Exercise 9.3.8. Calculate the determinant of both $D(T(2,3,4))$ and $N(T(2,3,4))$, and compare them with the continued fraction for $[4,3,2]$. (This problem explains why we use the terms *numerator* and *denominator* of a rational tangle.)

We know that the sum $A + B$, of rational tangles A and B, is not always a rational tangle. In fact, $D(A + B)$ need not necessarily be a 2-bridge knot (or link). However, the numerator $N(A+B)$ *is* a 2-bridge knot, or link (why is this?). Further, the following theorem explicitly calculates the determinant of $N(A + B)$.

Theorem 9.3.5.

Suppose that A and B are rational tangles with fractions $\frac{p}{q}$ and $\frac{r}{s}$, respectively. If we further suppose that the 2-bridge knot (or link) $N(A + B)$ is of type (α, β), then $\alpha = |ps + qr|$. [In this case α is the determinant of $N(A + B)$.]

Since a detailed proof of the above theorem is given in Ernst and Sumners [ES2], we refer the reader to that source. It is also possible to determine β; however, since its form is not as neat and it will not shed any substantial insight in what follows, we shall not discuss it further. The value of α, on the other hand, is expressed by a fraction $\frac{p}{q} + \frac{r}{s}$ that is not reduced is its numerator.

We shall verify that the above theorem is plausible by working through the next example.

Example 9.3.2. The fraction of $T(2,3,4)$ is $[4, 3, 2] = \frac{30}{7}$, while the fraction of $T(-3, -2)$ is $[-2, -3] = \frac{-7}{3}$. Therefore, the determinant of the numerator of $T(2, 3, 4) + T(-3, -2)$ is 41, since

$$\frac{30}{7} - \frac{7}{3} = \frac{90 - 49}{21} = \frac{41}{21}.$$

On the other hand, if we look at the regular diagrams, see Figure 9.3.9, we have that

$$N\big(T(2, 3, 4) + T(-3, -2)\big) = C(-2, -3, -1, -1, -2).$$

This knot is a 2-bridge knot of type $(41, -18)$.

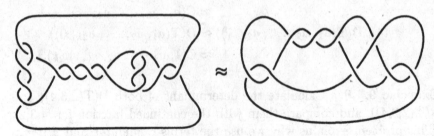

Figure 9.3.9

Exercise 9.3.9. Prove that if A is a rational tangle, then there always exists a rational tangle B such that $N(A+B)$ is a trivial knot. Further, find B in the case when $A = T(2,3,4)$.

2-bridge knots (or links) cannot be distinguished by means of their Alexander polynomials. For example, the 2-bridge knots of type (15,4) and (15,7) have both the same Alexander polynomial, namely, $4t^{-1} - 7 + 4t$, but they are not equivalent. Since 2-bridge knots (or links) are completely determined by their type (α, β), one would expect in theory all their invariants can be determined by α and β. In fact, the Alexander polynomial can be expressed via the continued fraction expansion $[a_1, a_2, \ldots, a_n]$ of $\frac{\alpha}{\beta}$. However, the actual method is quite complicated, cf. [BZ*]; so to avoid a turgid explanation, we shall restrict ourselves to showing how we can express the signature via the expansion.

Suppose K is a 2-bridge *knot* of type (α, β). It is sufficient (why?) to consider only the case $0 < \beta < \alpha$. We can also assume that β is odd [cf. Exercise 9.3.11(1)]. So, consider the sequence $\{0, \beta, 2\beta, \ldots, (\alpha - 1)\beta\}$, and divide each element by 2α to obtain a remainder r, $-\alpha < r < \alpha$, and these we may write down as the following sequence:

$$\{0, r_1, r_2, \ldots, r_{\alpha-1}\} \tag{9.3.1}$$

where $r_i \neq 0$, $i = 1, 2, \ldots, \alpha - 1$. Then the following theorem holds:

Theorem 9.3.6.
The signature, $\sigma(K)$, of K is equal to the number of positive entries minus the number of negative entries in the sequence of remainders given in (9.3.1).

Example 9.3.3. Suppose $(\alpha, \beta) = (7, 3)$. So, since
$$\{k\beta\} = \{0, 3, 6, 9, 12, 15, 18\},$$

we have that

$$\{r_k\} = \{0, 3, 6, -5, -2, 1, 4\};$$

so $\sigma(K) = 4 - 2 = 2$.

Of the classical invariants, the unknotting number of a 2-bridge knot is the only one that has yet to be completely decided.

Exercise 9.3.10. Determine the signature of the 2-bridge knots of type (15,4) and (25,7).

Exercise 9.3.11. (1) Show that for a 2-bridge knot (or link) of type (α, β), we can always take β to be an odd integer.

(2) Suppose L is a 2-bridge link of type $(2\alpha, \beta)$.

(i) Show that each component K_1 and K_2 of L is a trivial knot.

(ii) Show that the knot obtained from K_1 in the 2-fold cyclic covering space M ($\approx S^3$) of S^3 branched along K_2 is a 2-bridge knot of type (α, β).

(3) Show that a 2-bridge knot (or link) always has period 2. [Hint: Use (2)(ii).]

(Unoriented) 2-bridge knots (or links) can be characterized by their graphs (cf. Chapter 2, Section 3). We know that a 2-bridge knot (or link) has a standard diagram D given in Figure 9.3.7. By colouring the unbounded domain black, we obtain a signed graph G(D) of the form shown in Figure 9.3.10(a).

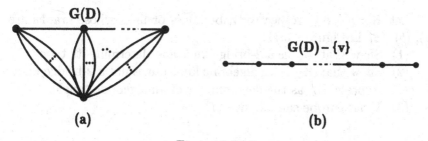

(a) **G(D)**   (b) **G(D) − {v}**

Figure 9.3.10

It is possible to find in G(D) a vertex v such that the subgraph $\hat{G} = G(D) - \{v\}$ that is obtained from G(D) by deleting v and all the edges incident to v, is a simple line segment, Figure 9.3.10(b). The converse is also true, and this is more formally expressed by the following proposition.

Proposition 9.3.7.

An (unoriented) knot (or link) K is a 2-bridge knot (or link) if and only if K has a regular diagram D with the following property:

The link graph G(D) (defined in Chapter 2, Section 3) has a vertex v such that G(D) − {v} is a simple line segment   (9.3.2)
(with vertices on it).

If condition (9.3.2) is slightly generalized to

G(D) has a vertex v such that G(D) − {v} is a tree,   (9.3.3)

then we can define a new class of knots (or links). A knot (or link) in this class is a called an *algebraic* link.

Exercise 9.3.12. Consider the three positive graphs in Figure 9.3.11.

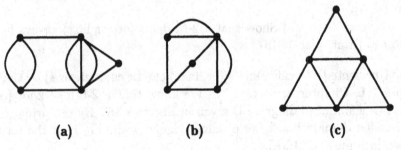

    (a)   **(b)**   **(c)**

Figure 9.3.11

Let K_a, K_b, K_c, respectively, be knots or links whose graphs are (a), (b), (c) in Figure 9.3.11.
    (1) Show that K_a is a 2-bridge knot and determine its type.
    (2) Show that K_b is an algebraic knot but is not a 2-bridge knot. Express K_b as the denominator of an algebraic tangle.
    (3) What can be said about K_c?

§4 Oriented 2-bridge knots

The bridge number of a knot or link, we know, does not depend on the orientation. Since we did not assign an orientation to the trivial tangle that we used originally to create the 2-bridge knots, these 2-bridge knots and links have no original orientation. Therefore, in order to investigate further the problem of equivalence with orientation of oriented knots, Theorem 9.3.3 is not sufficient. So in this final section we would like to look at the classification of oriented 2-bridge knots and links.

In the case of 2-bridge *knots,* there is, in fact, no problem to consider. The reason is that a 2-bridge knot is equivalent with orientation to the original knot, even if we reverse the orientation.

Exercise 9.4.1.   Show that a 2-bridge knot is invertible.

Suppose L is a 2-bridge link of type (α, β), with α even and β odd. Firstly, as before, let us determine the continued fraction expansion for $\frac{\alpha}{\beta}$, and hence construct the *unoriented* 2-bridge link $C(a_1, a_2, \ldots, a_{2k+1})$. Then to this link we may assign an orientation in the manner of Figure 9.4.1, making it an (standard) oriented 2-bridge link of type (α, β).

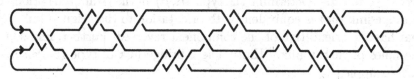

Figure 9.4.1

Exercise 9.4.2.   (1) Show that we can always assign an orientation to a 2-bridge link in the manner of Figure 9.4.1.

(2) Show that the 2-bridge link obtained by reversing the orientation on *each* of the two components of the (standard) oriented 2-bridge link is equivalent with orientation to the original link. Therefore, 2-bridge links are invertible.

If we reverse the orientation of *only* one component of a (standard) oriented 2-bridge link, L, of type (α, β), then, in general, it is not equivalent with orientation to L.

Exercise 9.4.3.   Show that the link obtained by reversing the orientation of one component of the (standard) oriented 2-bridge link of type $(4,1)$ is equivalent with orientation to the (standard) oriented 2-bridge link of type $(4, -3)$. Consider a similar question in conjunction with the 2-bridge links of type $(10,3)$ and $(10, -7)$.

The above problem suggests there might exist some sort of equivalence relations, with respect to orientation, for oriented 2-bridge links. The next theorem actually clarifies the matter.

Theorem 9.4.1 [Sc3].

(1) Suppose we reverse the orientation of one component of an (standard) oriented 2-bridge link of type (α, β), then the (standard) oriented 2-bridge link obtained is

     (i) *of type* $(\alpha, \beta - \alpha)$ *if* $\beta > 0$;

     (ii) *of type* $(\alpha, \beta + \alpha)$ *if* $\beta < 0$.

(2) Two (standard) oriented 2-bridge links of type (α, β) and (α', β') are equivalent with orientation if one of the following cases hold; further, these are sufficient conditions for them to be equivalent.

     (i) $\alpha = \alpha'$, $\quad \beta \equiv \beta' \pmod{2\alpha}$

or

     (ii) $\alpha = \alpha'$, $\quad \beta\beta' \equiv 1 \pmod{2\alpha}$.

Exercise 9.4.4. Show Theorem 9.4.1(2) is in the case of a 2-bridge knot equivalent to Theorem 9.3.3.

Exercise 9.4.5. Determine the type (α, β) of the (standard) oriented 2-bridge link that is equivalent with orientation to the original link but that has the orientation of one component reversed. Further, find some examples of such a link. [Hint: The 2-bridge link of type (8,3) is one such example.]

Exercise 9.4.6. Determine the formula that expresses the linking number of the (standard) oriented 2-bridge link of type (α, β). Using this formula, calculate the linking number of the (standard) oriented 2-bridge links of type (8,3) and (16,5).

Exercise 9.4.7. We can, by means of Theorem 9.3.6, calculate the signature of a (standard) oriented 2-bridge link. Using this theorem, determine the signature, $\sigma(L)$, of the (standard) oriented 2-bridge link of type (16,5), and the signature, $\sigma(L')$, of the 2-bridge link, L', obtained by reversing the orientation of one component. Finally, confirm the validity of (6.4.7) for this link.

The Theory of Braids

At the beginning of the 1930s, as a means of studying knots, E. Artin introduced a concept of a (mathematical) braid(s). This remarkable insight itself was not sufficient to sustain research in this area, and so it slowly began to wither. However, in the 1950s this concept of braids was found to have applications in other fields, and this gave fresh impetus to the study of braids, rekindling research in this area. The iridescent hue of this concept flowering into full bloom and activity occurred in 1984, when V. Jones put into action with inordinate success the original aim of Artin, i.e., the application of braids to knot theory.

In this chapter our intention is to introduce certain necessary aspects of the theory of braids that will prove useful when we explain recent developments in knot theory in the subsequent chapters.

§1 Braids

An n-braid is a very particular example of an (n, n)-tangle. On the top and base of a cube, B, mark out n points, $A_1, A_2, \ldots, A_n$ and $A_1{'}, A_2{'}, \ldots, A_n{'}$, respectively. These points may be arbitrarily placed, however, for the sake of neatness, we shall express them in terms of specific co-ordinates.

Firstly, the co-ordinates for B in $\mathbf{R}^3$ are

$$B = \{(x, y, z) \mid 0 \le x, y, z \le 1\}.$$

Let us choose $A_1, \ldots, A_n, A_1{'}, \ldots, A_n{'}$ as follows

$$A_1 = (\frac{1}{2}, \frac{1}{n+1}, 1), \quad \ldots \quad, A_n = (\frac{1}{2}, \frac{n}{n+1}, 1)$$
$$A_1' = (\frac{1}{2}, \frac{1}{n+1}, 0), \quad \ldots \quad, A_n' = (\frac{1}{2}, \frac{n}{n+1}, 0).$$

By construction each A_i' is directly below the corresponding A_i, Figure 10.1.1.

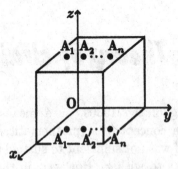

Figure 10.1.1

Now, join the $A_1, A_2, \ldots, A_n$ to $A_1{'}, A_2{'}, \ldots, A_n{'}$ by means of n curves (again, to be precise they should be polygonal arcs) in B. As usual, they are joined in such a way that these curves (including the endpoints) do not mutually intersect each other. It is *not* necessary to join A_i to A'_i, but we *cannot* join A_i to some A_j. We will call these polygonal arcs *strings*.

Suppose, now, that we divide the cube into two parts by an arbitrary plane E that is parallel to the base of the cube B. Then, if E intersects each string (polygonal arc) at one and only one point, we say that these n strings in B are an *n-braid*.

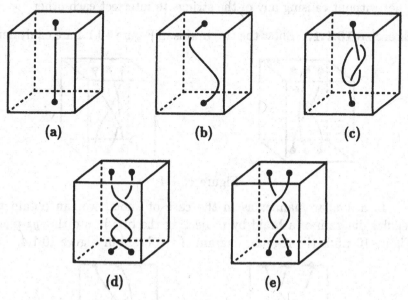

Figure 10.1.2

Example 10.1.1. Figure 10.1.2(a) and (b) are both examples of 1-braids. Figure 10.1.2(c), however, is *not* a (1-)braid. Figure 10.1.2(d) and (e) are typical examples of 2-braids.

If, given two n-braids in a cube, we can, by performing the elementary knot moves on these strings, transform one to the other, then we say that these two n-braids are equivalent (or equal).

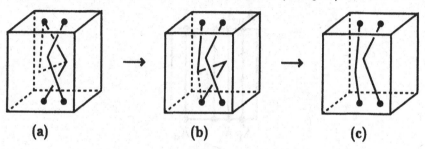

Figure 10.1.3

In the process of applying the elementary knot moves, it is perfectly possible that at some stage within the cube we obtain something that does not conform to our definition of a braid, Figure 10.1.3. Any 1-braid, therefore, is equivalent to the one in Figure 10.1.2(a).

Intuitively, two braids (in a cube), whose endpoints we keep fixed, can be said to be equivalent if we can continuously deform one to the other without causing any of the strings to intersect each other.

Exercise 10.1.1.   Show the two braids in Figure 10.1.4 are equivalent.

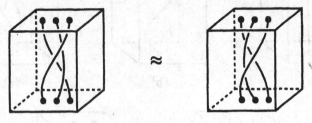

Figure 10.1.4

In a similar manner as in the case of knots, we can obtain the regular diagram of a braid by projecting the braid onto the yz-plane. Figure 10.1.5 is the regular diagram of the braid in Figure 10.1.4.

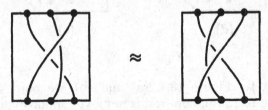

Figure 10.1.5

By connecting A_1 to A_1', A_2 to A_2', ..., A_n to A_n' by n line segments, Figure 10.1.6, we can form a special type of braid. In keeping with our previous nomenclature, we shall call this the *trivial n-braid*.

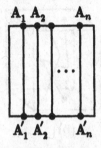

Figure 10.1.6

Suppose that a n-braid α has its strings connected as follows: A_1 to A'_{i_1}, A_2 to A'_{i_2}, ..., A_n to A'_{i_n}. Then we can assign to α a permutation,

$$\begin{pmatrix} 1 & 2 & \cdots & n \\ i_1 & i_2 & \cdots & i_n \end{pmatrix}.$$

We call this permutation the *braid permutation*. The trivial braid corresponds to the identity permutation,

$$\begin{pmatrix} 1 & 2 & \cdots & n \\ 1 & 2 & \cdots & n \end{pmatrix}.$$

Example 10.1.2. The braid permutation for Figure 10.1.2(d) is

$$\begin{pmatrix} 1 & 2 \\ 2 & 1 \end{pmatrix} = (1 \;\; 2),$$

while the braid permutation for Figure 10.1.4 is

$$\begin{pmatrix} 1 & 2 & 3 \\ 3 & 2 & 1 \end{pmatrix} = (1 \;\; 3).$$

Since if two braids are equivalent, it follows that their braid permutations are equal; so the braid permutation is a braid invariant. This invariant is not a number or polynomial, as we have used before, but still a mathematical concept, namely, a permutation. In fact, it is by far the simplest braid invariant.

§2 The braid group

Suppose that B_n is the set of all n-braids (to be more precise, all the equivalence classes of these braids). For two elements in B_n, i.e., for two n-braids α and β, it is possible to define a *product* for two n-braids α and β. Firstly, glue the base of the cube that contains α to the top face of the cube that contains β. The gluing together of the two cubes produces a rectangular solid in which there exists a braid that has been created from the vertical juxtaposition of α and β, Figure 10.2.1. (Obviously, we can recover a cube by shrinking the rectangular solid in half.)

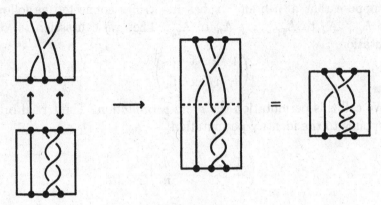

Figure 10.2.1

This braid is called the product of α and β, and is denoted by $\alpha\beta$. Similarly, we may define the product, $\beta\alpha$, of β and α. In general, it is not true that $\alpha\beta = \beta\alpha$, i.e., $\alpha\beta$ and $\beta\alpha$ need not be equivalent braids.

Exercise 10.2.1. Show the two products $\alpha\beta$ and $\beta\alpha$, of the α, β in Figure 10.2.2(a) and (b), are not equivalent. (Hint: Consider the permutations of $\alpha\beta$ and $\beta\alpha$.)

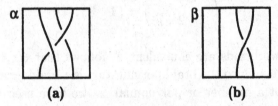

(a)   (b)

Figure 10.2.2

Although not necessarily commutative, braids are associative, i.e.,

$$(\alpha\beta)\gamma = \alpha(\beta\gamma).$$

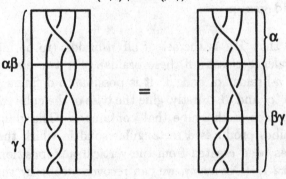

Figure 10.2.3

So far we have described a set B_n, a product in this set, and also that associativity holds in the set. Therefore, the natural question to consider is, Can we make B_n a group under the action of the product? In order to show this, we must find a unit element and an inverse element. The unit e is simply the trivial braid, Figure 10.1.6. It follows readily from Figure 10.2.4 that, irrespective of the braid, α, $\alpha e = \alpha$, and similarly, $e\alpha = \alpha$.

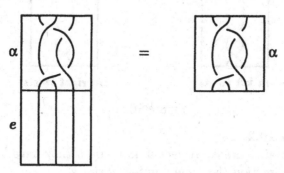

Figure 10.2.4

In order to find an inverse for an arbitrary α, let us consider the mirror image, α^*, of α. If we consider the base of the cube to be a mirror, then the mirror image is the image of α reflected in this mirror.

Exercise 10.2.2. Show that $\alpha^*\alpha$ and $\alpha\alpha^*$ are equivalent to the trivial braid e; see also Figure 10.2.5.

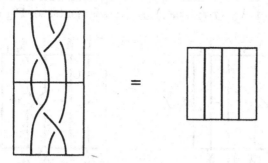

Figure 10.2.5

Since this exercise can be solved, we may write that $\alpha\alpha^* = e$ and $\alpha^*\alpha = e$. Therefore, we now have all the essentials for B_n to be a group. This group is called the *n-braid group*. The inverse element, α^*, of α we shall denote by α^{-1}.

Let us delve a bit further into the structure of these groups. Firstly, since the 1-braid group, B_1, contains only one element, namely, the trivial braid, $B_1 = \{e\}$. The elements of B_2 are equivalent to the two types of braids drawn in Figure 10.2.6. We have, in fact,

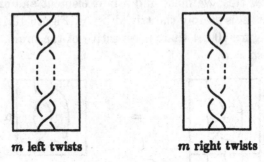

m left twists   m right twists

Figure 10.2.6

Proposition 10.2.1.

Two 2-braids are equivalent if and only if they have been twisted in the same direction the same number of times.

(For a proof see Exercise 10.3.4.) Therefore, B_2 has an infinite number of non-equivalent elements; i.e., it is a group of infinite order. For $n \geq 2$, every B_n is a group of infinite order; however, there exists a very easy way of actually writing a general element in one of these groups.

Among the n-braids, we can create certain specific n-braids by connecting A_i to A'_{i+1} and A_{i+1} to A'_i, and then connecting the remaining A_j and A'_j $(j \neq i, i+1)$ by line segments, see Figure 10.2.7(a).

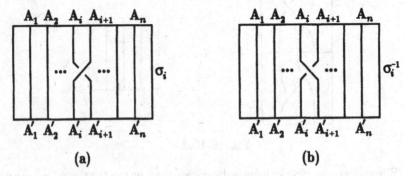

(a)   (b)

Figure 10.2.7

We shall denote these types of n-braids by σ_i. In this way we can create $n-1$ special n-braids $\sigma_1, \sigma_2, \ldots, \sigma_{n-1}$. In Figure 10.2.7(b) we have also drawn the inverse element of σ_i, the n-braids σ_i^{-1}. We may now use these elements to express any element in the braid group. For example, in Figure 10.2.8 we have drawn the braid $\alpha = \sigma_2\sigma_1\sigma_2^{-1}\sigma_3\sigma_1$.

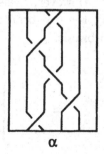

α

Figure 10.2.8

Conversely, to express any element of B_n in terms of these σ_i and σ_i^{-1}, first we divide the regular diagram of a braid by lines parallel to the bottom edge, so that in each rectangle we have only *one* crossing point. (If two crossing points are at the same level, then by shifting one slightly upwards and the other slightly downwards, we can eliminate the problem of having two crossing points at the same level.)

In each of these rectangles, by construction, we have a braid that is of the form σ_i or σ_i^{-1}. By definition of the product of braids, we can decompose β into the product of these σ_i and σ_i^{-1}. For example, the braid β in Figure 10.2.9 is $\sigma_3^{-1}\sigma_1\sigma_2\sigma_3\sigma_2^{-1}$.

β

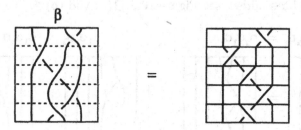

$=$

Figure 10.2.9

Therefore, given any braid, we can express it as the finite product of the σ_i and σ_i^{-1}. For this reason, the braids $\sigma_1, \sigma_2, \ldots, \sigma_{n-1}$ are said to *generate* the braid group B_n, and so we call $\sigma_1, \sigma_2, \ldots, \sigma_{n-1}$ the *generators* of B_n. For example, since any 2-braid may be written as σ_1^m or σ_1^{-m}, where of course $m \geq 0$ and

$$\sigma_1^m = \underbrace{\sigma_1 \cdots \sigma_1}_{m \text{ times}} \quad \text{and} \quad \sigma_1^{-m} = \underbrace{\sigma_1^{-1} \cdots \sigma_1^{-1}}_{m \text{ times}},$$

B_2 is generated by a single element, σ_1.

From the above, we have a way of describing algebraically a braid as a product of σ_i and σ_i^{-1}. However, these algebraic representations are not unique. For example, the two braids $\sigma_1\sigma_3$ and $\sigma_3\sigma_1$, in Figure 10.2.10, are equivalent 4-braids.

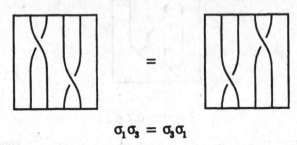

$$\sigma_1\sigma_3 = \sigma_3\sigma_1$$

Figure 10.2.10

Therefore, in the 4-braid group, B_4, the equation $\sigma_1\sigma_3 = \sigma_3\sigma_1$ holds. Further, since $\sigma_1\sigma_2\sigma_1$ and $\sigma_2\sigma_1\sigma_2$ are equivalent 3-braids (cf. Figure 10.1.4 and 10.1.5), the following relation holds:

$$\sigma_1\sigma_2\sigma_1 = \sigma_2\sigma_1\sigma_2.$$

This equality holds even if this braid is considered as a general n-braid $(n \geq 3)$, i.e., the regular diagram has a few extra non-intersecting lines added, see Figure 10.2.11(a) and (b).

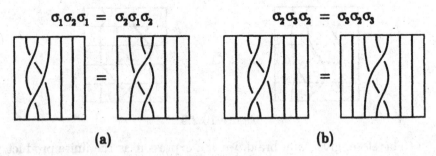

(a)   (b)

Figure 10.2.11

These equalities are called *(braid) relations* of the braid group. In fact, if two n-braids are equivalent, then we can change one to other by using these equalities several times; an example is given a bit later in Example 10.2.1.

A fundamental result on the braid group B_n is that it only has the following two type of relations called the fundamental relations:

$$(1) \quad \sigma_i \sigma_j = \sigma_j \sigma_i \qquad\qquad (|i - j| \geq 2);$$
$$(2) \quad \sigma_i \sigma_{i+1} \sigma_i = \sigma_{i+1} \sigma_i \sigma_{i+1} \quad (i = 1, 2, \ldots, n - 2). \qquad (10.2.1)$$

(Of course, there exist trivial relations, namely, $\sigma_i \sigma_i^{-1} = e$ and $\sigma_i \sigma_j = \sigma_i \sigma_j$; however, we will not consider these as *bona fide* relations, and so we will ignore them.)

Collecting together the various relations we have discussed this far, we may write B_n in terms of its generators $\sigma_1, \sigma_2, \ldots, \sigma_{n-1}$ and the these fundamental relations,

$$B_n = \left(\sigma_1, \sigma_2, \ldots, \sigma_{n-1} \,\middle|\, \begin{array}{ll} \sigma_i \sigma_j = \sigma_j \sigma_i & (|i - j| \geq 2) \\ \sigma_i \sigma_{i+1} \sigma_i = \sigma_{i+1} \sigma_i \sigma_{i+1} & (i = 1, 2, \ldots, n - 2). \end{array} \right)$$

where the right-hand side is said to be a *presentation* of B_n.

For example,

$$B_1 = (\sigma_1 \mid \text{\textemdash\textemdash} \,),$$
$$B_2 = (\sigma_1, \sigma_2 \mid \sigma_1 \sigma_2 \sigma_1 = \sigma_2 \sigma_1 \sigma_2 \,),$$
$$B_3 = (\sigma_1, \sigma_2, \sigma_3 \mid \sigma_1 \sigma_3 = \sigma_3 \sigma_1, \; \sigma_1 \sigma_2 \sigma_1 = \sigma_2 \sigma_1 \sigma_2, \; \sigma_2 \sigma_3 \sigma_2 = \sigma_3 \sigma_2 \sigma_3 \,).$$

Since, except for the trivial relation $\sigma_1 \sigma_1^{-1} = e$, B_1 does not have any relations, we denote this lack of relations by \textemdash.

Exercise 10.2.3. Determine presentations for the braid groups B_5 and B_6. (Hint: B_5 has 6 fundamental relations, and B_6 has 10 fundamental relations.)

Exercise 10.2.4. Show that the following equalities hold for any B_n $(n \geq 3)$:

$$(1) \quad \sigma_1 \sigma_2 \sigma_1^{-1} = \sigma_2^{-1} \sigma_1 \sigma_2$$
$$(2) \quad \sigma_2 \sigma_1^{-1} \sigma_2^{-1} = \sigma_1^{-1} \sigma_2^{-1} \sigma_1.$$

Example 10.2.1. It follows from Figure 10.2.12 that the two elements, α and β, of B_3 given by

$$\alpha = \sigma_1 \sigma_2^2 \sigma_1^{-1} \sigma_2^{-1} \sigma_1^2 \sigma_2 \quad \text{and} \quad \beta = \sigma_2^{-1} \sigma_1^4 \sigma_2.$$

are equal.

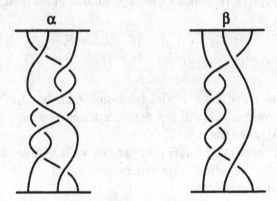

Figure 10.2.12

Let us show this, however, by transforming α into β by applying various braid relations.

We intend to use the relations in Exercise 10.2.4, rather than the fundamental ones, since this will make the calculations much more transparent. Hence, we may change the part $\sigma_2\sigma_1^{-1}\sigma_2^{-1}$ in α to $\sigma_1^{-1}\sigma_2^{-1}\sigma_1$, and so

$$\alpha = \sigma_1\sigma_2(\sigma_1^{-1}\sigma_2^{-1}\sigma_1)\sigma_1^2\sigma_2 = \boxed{\sigma_1\sigma_2\sigma_1^{-1}}\,\sigma_2^{-1}\sigma_1^3\sigma_2.$$

In Exercise 10.2.4 (1) we showed that $\sigma_1\sigma_2\sigma_1^{-1} = \sigma_2^{-1}\sigma_1\sigma_2$. Using this equality we can change the portion inside $\boxed{}$ to $\sigma_2^{-1}\sigma_1\sigma_2$, thus obtaining

$$\alpha = \sigma_2^{-1}\sigma_1\sigma_2\sigma_2^{-1}\sigma_1^3\sigma_2 = \sigma_2^{-1}\sigma_1^4\sigma_2 = \beta.$$

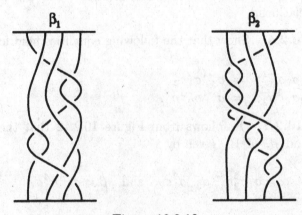

Figure 10.2.13

Exercise 10.2.5. Show by means of the regular diagrams that the two elements β_1 and β_2 in Figure 10.2.13, of B_4, are equivalent. By expressing β_1 and β_2, respectively, in terms of σ_i and σ_i^{-1}, change β_1 to β_2 by means of the braid relations of B_4 (and the equalities that are derived from these braid relations).

§3 Knots and braids

Let us connect, by a set of parallel arcs that lie outside the square, the points $A_1, A_2, \ldots, A_n$ on the top of a rectangular diagram of a braid α to the points $A_1', A_2', \ldots, A_n'$, respectively, on the bottom of the same diagram; see, for example, Figure 10.3.1.

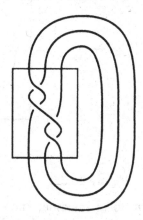

Figure 10.3.1

Then in a natural way we form a regular diagram of knot or link from a braid. A knot that has been created in this way is said to be a knot (or link), K, *created* (or *formed*) from the braid α. Conversely, we can say that K is the *closed braid* (or the *closure* of α). Usually we assign to each string an orientation that starts from A_i and then moves downwards along the corresponding string in the cube. Hence, from a braid α we can form an oriented knot (or link) K. Conversely, given an oriented knot we can change it suitably so that it becomes an oriented closed braid. This is encapsulated in the next theorem.

Theorem 10.3.1 (Alexander's theorem).

Given an arbitrary (oriented) knot (or link), then it is equivalent (with orientation) to a knot (or link) that has been formed from a braid.

Proof

(In Example 10.3.1 and its accompanying diagrams, we show explicitly how to form a braid from a knot, so we refer the reader to this example as template for the proof of this theorem.)

Suppose D is an oriented regular diagram of a knot K. Firstly, cut D at a point (but not a crossing point) P_0, and then pull the loose ends apart so that we now have a (regular diagram of a) $(1,1)$-tangle T, Figure 10.3.2. We shall show that we can change this tangle into a braid α. The knot, in a sense induced as described previously from this braid, is equivalent to K.

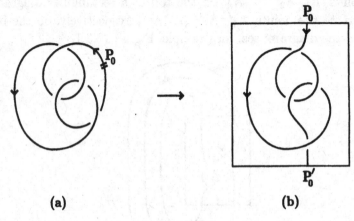

(a)	(b)

Figure 10.3.2

If the tangle T has m local maxima, then it also has m local minima. In the case $m = 0$, as previously noted, T is a 1-braid and so no proof is required.

So suppose $m > 0$, then there exists an arc $\overset{\frown}{ab}$ in T, which we may say is *rising upwards*, connecting a local minimum a to a local maximum b, Figure 10.3.3(a).

Further, we may assume that $\overset{\frown}{ab}$ intersects with the other parts of the tangle at n places. Let us now mark $n + 1$ points on $\overset{\frown}{ab}$, i.e., $a = a_0, a_1, \ldots, a_n = b$, such that the arc $\overset{\frown}{a_i a_{i+1}}$ intersects *only* one other part of the tangle, see Figure 10.3.3(a). Next, replace the arc $\overset{\frown}{a_0 a_1}$ by the much larger arc $\overset{\frown}{a_0 P_1' P_1 a_1}$. The (large) arc $\overset{\frown}{P_1 P_1'}$ lies outside the tangle T, and the arcs $\overset{\frown}{a_0 P_1'}$ and $\overset{\frown}{a_1 P_1}$ are selected in such a way that if $\overset{\frown}{a_0 a_1}$ passes over (or under) the other segment, then they ($\overparen{a_0P_1'}$ and $\overparen{a_1P_1}$) also pass over (or under) all the other segments. The result of the above manipulations is a $(2,2)$-tangle, see Figure 10.3.3(b).

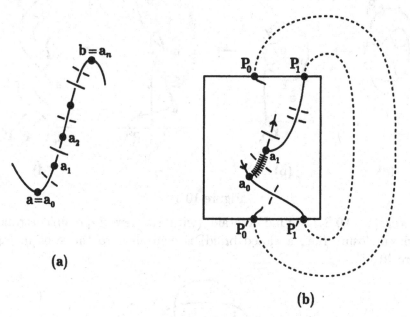

(a)

(b)

Figure 10.3.3

It follows immediately that the oriented knot obtained by joining (outside the square) the four endpoints of this $(2,2)$-tangle by curves is equivalent to the original knot. [In this tangle, a ($= a_0$) is no longer a local minimum, and a_1 is a new local minimum.] By using the same methods as above, with regard to the arcs $\overparen{a_1a_2}$, $\overparen{a_2a_3}$, ..., $\overparen{a_{n-1}a_n}$ ($=$ $\overparen{a_{n-1}b}$), we will eventually form a $(n+1, n+1)$-tangle T that does not have a local minimum and maximum at a and b respectively, and further has at most $m - 1$ local maxima and minima. Continuing this procedure, we shall finally create a tangle that has no local maxima and minima. This tangle is our required braid. The same proof will also work for a link.

■

Example 10.3.1. Figure 10.3.4(a) $\sim$ (d) shows how the procedure of the above proof works in practice on a regular diagram of a knot. Confirm that the knot formed from this braid is equivalent to the original knot.

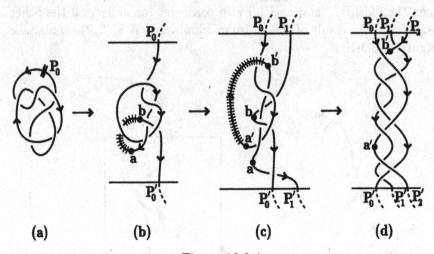

Figure 10.3.4

Exercise 10.3.1. Find a braid (with the fewest possible strings) whose closure (i.e., a closed braid) is equivalent to the knot in Figure 10.3.5.

Figure 10.3.5

Now, if two braids are equivalent, their knots (obtained by closing the braid), it goes without saying, are also equivalent. *Caveat lector*, it is also possible to obtain equivalent knots from the closure of *nonequivalent* braids. For example, the braids in Figure 10.3.6 are non-equivalent, but their closures are equivalent, to the trivial knot.

Figure 10.3.6

Exercise 10.3.2. Show that the closure of the 2-braid σ_1^{-1}, the 3-braids $\sigma_1^{-1}\sigma_2$ and $\sigma_1\sigma_2^{-1}$, and the 4-braids $\sigma_1\sigma_2\sigma_3^{-1}$ and $\sigma_1\sigma_2^{-1}\sigma_3^{-1}$ are all equivalent to the trivial knot.

Therefore, if we wish to apply braid theory to knot theory, we must first of all explain clearly "how to restrict the braids from *which* we can form equivalent knots." To this end, we shall introduce between two braids the concept of M-equivalence.

Definition 10.3.1. Suppose that B_∞ is the union of the groups $B_1, B_2, \ldots, B_n, \ldots$, i.e., $B_\infty = \bigcup_{k \geq 1} B_k$. We may perform the following two operations in B_∞; the operations are called *Markov moves*:

(1) If β is an element of the braid group B_n (i.e., β is an n-braid), then M_1 is the operation that transforms β into the n-braid $\gamma\beta\gamma^{-1}$, where γ is some element of B_n, see Figure 10.3.7(a). The element $\gamma\beta\gamma^{-1}$ is the *conjugate* of β.

(2) M_2 is the operation that transforms a n-braid, β, into either of the two $(n+1)$-braids $\beta\sigma_n$ or $\beta\sigma_n^{-1}$, where σ_n is a generator of B_{n+1}, the $(n+1)$-braid group, see Figure 10.3.7(b).

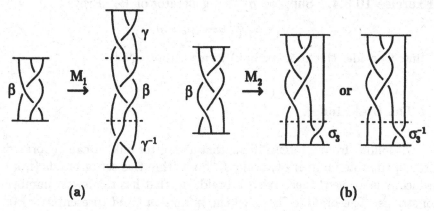

(a)   **(b)**

Figure 10.3.7

Example 10.3.2. Figure 10.3.7(a) shows how $\beta = \sigma_2\sigma_1^{-1}\sigma_2$, an element of B_3, changes when M_1 is applied; namely, β becomes $\gamma\beta\gamma^{-1}$, where in this case $\gamma = \sigma_2\sigma_1^{-1}$. Figure 10.3.7(b) shows how $\beta = \sigma_2\sigma_1^{-1}\sigma_2\sigma_1^{-1}$, an element of B_3, changes when we apply M_2; i.e., we obtain the element $\beta\sigma_3$ or $\beta\sigma_3^{-1}$ of B_4.

Definition 10.3.2. Suppose that α and β are elements of B_∞. If we can transform α into β by performing the Markov moves M_1, M_2, and their inverses M_1^{-1}, M_2^{-1} a finite number of times, then α is said to be *Markov equivalent* (M-equivalent) to β and is denoted by $\alpha \underset{M}{\sim} \beta$. If $\alpha \underset{M}{\sim} \beta$, then since $\beta \underset{M}{\sim} \alpha$, α and β are said to be Markov equivalent.

The following theorem shows that Markov equivalence is the fundamental concept that connects a knot to a braid.

Theorem 10.3.2 (Markov's theorem).

Suppose that K_1 and K_2 are two oriented knots (or links), which can be formed from the braids β_1 and β_2, respectively. Then

$$K_1 \cong K_2 \iff \beta_1 \underset{M}{\sim} \beta_2.$$

Exercise 10.3.3. Confirm that if $\beta_1 \underset{M}{\sim} \beta_2$ then $K_1 \cong K_2$ for the cases in Figure 10.3.7(a) and (b).

The above theorem was first announced by Markov in 1936, however an exemplary, complete proof appears in Birman [B*].

Exercise 10.3.4. Suppose σ_1 is a generator of B_2. Prove that

$$\sigma_1^m = \sigma_1^n \iff m = n.$$

[Hint: Consider the knot (or link) formed from σ_1^m.]

§4 The braid index

It follows from Exercise 10.3.2 that a knot (or link) K can be formed from a (infinite number of) braid(s). So, within this set of braids (from which K is formed) there exists a braid, α, that has the fewest number of strings. The braid α is called the *minimum braid* (presentation) of K, and the number of strings is said to be the braid index of K, and is denoted by b(K). (*Caveat lector*, the minimum braid presentation of K is not unique.) For example, the trivial knot has braid index 1, while, conversely, the knot(s) of braid index 1 is only the trivial knot.

Exercise 10.4.1. Show that b(K), the braid index of K, is an invariant of K.

The knots (or links) with braid index 2 are the elementary torus knots, i.e., only the torus knots of type $(n, 2)$, where $n \neq 0, \pm 1$.

The knots (or links) with braid index 3, in general, are difficult to list. In fact, it was only relatively recently that it was shown that certain types of oriented 2-bridge knots and links have braid index 3, cf. Chapter 11, Section 5. We should now add our customary remark: As yet no general algorithm has been found to calculate b(K).

Exercise 10.4.2. Determine the minimum braid presentation of the knot in Figure 10.3.5. (Hint: The braid index is 3.)

Exercise 10.4.3. Show the braid index of the 2-bridge knot of type (45,7) is 3.

If L is a link, then b(L) is related to the orientation of L. It is often the case that if we reverse the orientation of a *single* component, then the braid index will also change.

Exercise 10.4.4. Show that if we give the torus link $K_{4,2}$ two different orientations, then the respective braid indices are different.

In specific cases, for example, *torus knots*, as shown below, we can completely determine their braid index.

Suppose K is a knot and α is the minimum braid presentation of K. Then b(K) is the number of strings of α. In this case, since the regular diagram of K is obtained by joining b(K) semicircles (the closure strings) in parallel with the outside of the regular diagram of α (Figure 10.3.1), the regular diagram of K has exactly b(K) local maxima. Since the bridge number of K, br(K), is the minimum number of these maxima (Theorem 4.3.3), it follows that br(K) ≤ b(K).

Proposition 10.4.1.
Suppose br(K) *is the bridge number and* b(K) *is the braid index of a (oriented) knot (or link) K, then*

$$\mathrm{br}(K) \leq \mathrm{b}(K).$$

We know (Theorem 7.5.3) that the bridge number of the torus knot of type (q, r), where $q, r > 0$, $K(q, r)$, is

$$\mathrm{br}(K(q, r)) = \min(q, r).$$

But, since $K(q, r) = K(r, q)$, we may assume that $r < q$. Further, $K(q, r)$ can be formed from the r-braid (cf. Figure 7.1.7):

$$(\sigma_{r-1}\sigma_{r-2} \cdots \sigma_2\sigma_1)^q.$$

Therefore,

$$b(K(q,r)) \leq r = \mathrm{br}(K(q,r)).$$

So, combining this with Proposition 10.4.1, we obtain

$$b(K(q,r)) = \mathrm{br}(K(q,r)) = r \ (= \min \{q,r\}).$$

Proposition 10.4.2.
  If K *is a torus knot of type* (q,r), $q \neq 0 \neq r$, *then*

$$b(K) = \min \{|q|, |r|\}.$$

In general, determining the braid index is a difficult problem. However, recently, as will be explained in the next chapter, it has been shown that $b(K)$ is related to the degree of the skein polynomial. So, we can end this chapter optimistically and without our usual statement: "As yet no algorithm ... ." Also, due to the above-mentioned result, we can completely determine the braid index of 2-bridge knots or links; the exact value will be discussed in Chapter 11, Section 5.

The Jones Revolution

In 1984, after nearly half a century in which the main focus in knot theory was the knot invariants derived from the Seifert matrix, for example, the Alexander polynomial, the signature of a knot, *et cetera*, V. Jones announced the discovery of a new invariant. Instead of further propagating pure theory in knot theory, this new invariant and its subsequent offshoots unlocked connections to various applicable disciplines, some of which we will discuss in the subsequent chapters.

In the previous chapter we showed how to transform a braid into a knot and how to create a group from the braids. Therefore, we have the following correspondence:

$$\text{knot} \implies \text{braid} \implies \text{braid group, } B_n.$$

Suppose we can map the braid group B_n into some sort of algebraic system, say, A, whose structure we understand, for example, the group of invertible matrices, or more generally, an algebra such as a group ring in which the sum and product have been defined. The aim is to be able to represent an arbitrary knot by an element of A. (Of course, we must take care in our selection of A and the correspondence between the knots and A, since a situation may arise in which the correspondence assigns the same element to each knot.) It should be noted that such an approach is not exactly new; the Alexander polynomial can be obtained as an expression in a matrix ring with elements Laurent polynomials in a single indeterminate. However, an initial stumbling block is that the correspondence

$$\text{knot} \implies \text{braid}$$

is not in 1-1 correspondence. To be more precise, to a single knot we may assign an infinite number of braids. It is our good fortune, due to Markov's theorem (Theorem 10.3.2), that each knot corresponds to only one *M-equivalence class*. Therefore, when a braid α corresponds to a certain value, say, $\phi(\alpha)$, then if this value $\phi(\alpha)$ is the same for any other M-equivalent braid β, it follows that this $\phi(\alpha)$ is an invariant of the knot K_α formed from the braid α. So, from the first condition of M-equivalence ϕ must have the same value for α and $\gamma\alpha\gamma^{-1}$. A typical example of such a function is the trace of a square matrix (i.e., the sum of its diagonal elements). If we associate α with some square matrix, then if $\phi(\alpha)$ is the trace of this matrix, it follows immediately that $\phi(\alpha)$ is invariant under the Markov move M_1.

Now, if we want to represent the braid group by some algebraic system, A, it must have a similar structure to B_n. Further, A should have a simpler or more restricted algebraic structure than B_n; otherwise, we will probably not gain any further insight into B_n.

Jones, serendipitously, found that one of the (operator) algebras, which he was studying for other purposes, had a structure that resembled that of the braid group. By means of this insight, Jones was able to define a function that was invariant under both the Markov moves, M_1 and M_2. This function could, in fact, be written in terms of a complex number q. Following from this, it was possible to assign to each knot a complex number. To be more exact, to every knot it became possible to associate a Laurent polynomial in this complex number q (if we replace *q* by an indeterminate t, then we shall recover the usually defined Laurent polynomial). This (Laurent) polynomial, one of the new invariants, is now called the *Jones polynomial*.

Soon after Jones' announcement of the discovery of his polynomial, it became clear that this polynomial could be constructed using methods from other disciplines, for example, statistical mechanics, quantum groups, *et cetera*. Hence, knot theory was once again entwined with fields outside pure mathematics, generating a tremendous amount of interdisciplinary research and virtually spawning a whole new area of research.

In this chapter, our intention is to study the new invariants from the point of view of knot theory, explaining several of their fundamental properties. Also, in the final section we shall show an application of the Jones polynomial to knot theory itself, namely, solving a couple of the Tait conjectures, the original knot theory conjectures.

§1 The Jones polynomial

Let us begin by defining the Jones polynomial from the perspective of knot theory, rather than, say, operator algebras or quantum groups, which would require the introduction of a great deal of new notation and definitions, without significantly illuminating our further discussions.

Definition 11.1.1. Suppose K is an oriented knot (or link) and D is a (oriented) regular diagram for K. Then the Jones polynomial of K, $V_K(t)$, can be defined (uniquely) from the following two axioms. The polynomial itself is a Laurent polynomial in $\sqrt{t}$, i.e., it may have terms in which $\sqrt{t}$ has a negative exponent. [We assume $(\sqrt{t})^2 = t$.] The polynomial $V_K(t)$ is an invariant of K.

Axiom 1 : If K is the trivial knot, then $V_K(t) = 1$.

Axiom 2 : Suppose that D_+, D_-, D_0 are skein diagrams (cf. Figure 6.2.1), then the following skein relation holds.

$$\frac{1}{t}V_{D_+}(t) - tV_{D_-}(t) = \left(\sqrt{t} - \frac{1}{\sqrt{t}}\right)V_{D_0}(t). \tag{11.1.1}$$

The reader may at first sight mistake these axioms for those that define the Alexander polynomial, in particular, equations (11.1.1) and (6.2.1). In fact, it was only a number of years after the discovery of the

Jones polynomial that an extremely unexpected relationship was found between the Alexander polynomial and in a sense a "truncated" form of the Jones polynomial [MelM]; see also Chapter 15, Section 3. When the Jones polynomial was first announced, however, many people, not just mathematicians, were taken aback by the fact that such a seemingly incongruous change in the second axiom of the Alexander polynomial should have such a profound significance, and indeed it has been profound. For the proof that the Jones polynomial is defined uniquely by the above axioms, we refer the reader to Lickorish and Millett [LM3] and Jones [J1].

The algorithm to calculate the Jones polynomial is completely analogous to the one for the Alexander polynomial, which we explained in Chapter 6, Section 2; i.e., it is necessary to form a skein tree diagram. However, since the coefficients of the two skein relations *are* different, the calculation in the Jones case becomes a trifle difficult. The difficulty (or perhaps its strength) may immediately be seen if we write out the Jones polynomial as the sum of the Jones polynomials of the trivial μ-component links, O_μ, the result of using the skein tree diagram,

$$V_K(t) = f_1(t)V_O(t) + f_2(t)V_{OO}(t) + \; \cdots \; f_n(t)V_{O_n}(t).$$

In the Alexander case, a similar expression to that above is superfluous since we have already shown that the Alexander polynomial of O_μ for $\mu \geq 2$ is zero in Proposition 6.2.2, while in the Jones case and herein lies the fundamental difference, they are not zero. Therefore, our first step is to calculate the Jones polynomial of O_μ, $V_{O_\mu}(t)$.

Proposition 11.1.1.
For the trivial μ-component link O_μ,

$$V_{O_\mu}(t) = (-1)^{\mu-1}\left(\sqrt{t} + \frac{1}{\sqrt{t}}\right)^{\mu-1}. \qquad (11.1.2)$$

<u>Proof</u>
The proof will be by induction on μ. If $\mu = 1$, then this is just Axiom 1. So, let us assume for our induction hypothesis that the following holds:

$$V_{O_{\mu-1}}(t) = (-1)^{\mu-2}\left(\sqrt{t} + \frac{1}{\sqrt{t}}\right)^{\mu-2}.$$

If we now consider the skein diagram in Figure 6.2.5, then since $D_+ \cong D_- \cong O_{\mu-1}$ and $D_0 \cong O_\mu$, by the above induction hypothesis and the skein relation (11.1.1),

$$\frac{1}{t}(-1)^{\mu-2}\left(\sqrt{t}+\frac{1}{\sqrt{t}}\right)^{\mu-2} - t(-1)^{\mu-2}\left(\sqrt{t}+\frac{1}{\sqrt{t}}\right)^{\mu-2}$$

$$= \left(\sqrt{t}-\frac{1}{\sqrt{t}}\right)V_{O_\mu}(t).$$

Since the left-hand side of the above formula is

$$(-1)^{\mu-2}\left(\sqrt{t}+\frac{1}{\sqrt{t}}\right)^{\mu-2}\left(\frac{1}{t}-t\right)$$

$$= (-1)^{\mu-1}\left(\sqrt{t}+\frac{1}{\sqrt{t}}\right)^{\mu-2}\left(\sqrt{t}+\frac{1}{\sqrt{t}}\right)\left(\sqrt{t}-\frac{1}{\sqrt{t}}\right),$$

the required result (11.1.2) follows immediately. ∎

We now have the essentials to give an example of a calculation of the Jones polynomial. As in the Alexander polynomial case, let us write down the following equalities:

$$V_{D_+}(t) = t^2 V_{D_-}(t) + tz V_{D_0}(t)$$
$$V_{D_-}(t) = t^{-2} V_{D_+}(t) - t^{-1} z V_{D_0}(t),$$

where for clarity we have set $z = \left(\sqrt{t}-\frac{1}{\sqrt{t}}\right)$.

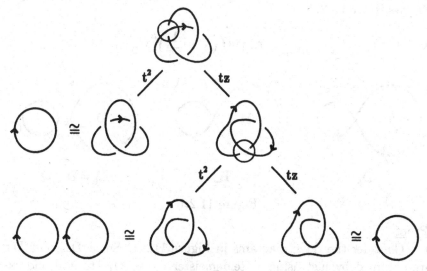

Figure 11.1.1

Example 11.1.1. In Figure 11.1.1 we have drawn the skein tree diagram for the calculation of the Jones polynomial of the right-hand trefoil knot (it is an interesting exercise for the reader to compare it with the one in Figure 6.2.6).

It follows from the skein tree diagram that

$$V_K(t) = t^2 V_O(t) + t^3 z V_{OO}(t) + t^2 z^2 V_O(t)$$
$$= t + t^3 - t^4.$$

Exercise 11.1.1. Calculate the Jones polynomial of the figure 8 knot, the knot in Figure 5.1.5(a), and the Whitehead link.

In Appendix (II) we list the Jones polynomial of each prime knot with up to 8 crossing points.

§2 The basic characteristics of the Jones polynomial

Let us denote by $X \amalg Y$ the union of two sets X and Y that have no points in common. For example, we may write the regular diagram of the 2-component trivial link as $O \amalg O$. Before discussing some of the characteristics of the Jones polynomial, let us first extend Proposition 11.1.1.

Proposition 11.2.1.

$$V_{D \amalg O_\mu} = (-1)^\mu \left(\sqrt{t} + \frac{1}{\sqrt{t}} \right)^\mu V_D(t)$$

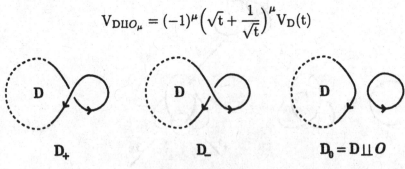

$$D_+ \qquad\qquad D_- \qquad\qquad D_0 = D \amalg O$$

Figure 11.2.1

Proof
    Consider the skein diagrams in Figure 11.2.1. Since D_+ and D_- are obtained by just using a Reidemeister move, Ω_1, to add one extra loop to D, $D_+ \cong D_- \cong D$. Thus substituting this information in

(11.1.1), namely, the skein relation is

$$\frac{1}{t}V_{D_+}(t) - tV_{D_-}(t) = \left(\sqrt{t} - \frac{1}{\sqrt{t}}\right)V_{D_0}(t),$$

then

$$V_{D_0}(t) = \frac{1}{\left(\sqrt{t} - \frac{1}{\sqrt{t}}\right)}\left(\frac{1}{t} - t\right)V_D(t) = -\left(\sqrt{t} + \frac{1}{\sqrt{t}}\right)V_D(t).$$

Hence, the case $\mu = 1$ has been proven. The general case follows by using mathematical induction, and since this is a straightforward exercise, we leave it to the reader.

∎

Theorem 11.2.2.
Suppose that $K_1 \# K_2$ is the (connected) sum of two knots (or links); then

$$V_{K_1 \# K_2}(t) = V_{K_1}(t)V_{K_2}(t).$$

Proof
Suppose D_1 and D_2 are, respectively, the regular diagrams of K_1 and K_2. Let us suppose, *temporarily*, that D_2 is invisible (for example, associate D_2 with a single black spot). So, ignoring D_2 let us create the skein tree diagram for D_1. Hence, from this we may write down the Jones polynomial of K_1,

$$V_{K_1}(t) = f_1(t)V_O(t) + f_2(t)V_{OO}(t) + \ \cdots \ + f_m(t)V_{O_m}(t). \quad (11.2.1)$$

To continue the calculation, we must now make D_2 visible again. By construction one component of each trivial link that makes up the endpoints of the skein tree diagram of D_1 has a black dot on it. This component, when we make D_2 visible, represents D_2. We may thus modify (11.2.1) to

$$V_{K_1 \# K_2}(t) = f_1(t)V_{D_2}(t) + f_2(t)V_{D_2 \sqcup O}(t)$$
$$+ f_3(t)V_{D_2 \sqcup O_2}(t) + \ \cdots \ + f_m(t)V_{D_2 \sqcup O_{m-1}}(t). \quad (11.2.2)$$

However, from Proposition 11.2.1 it follows that

$$V_{D_2 \sqcup O_k}(t) = (-1)^k \left(\sqrt{t} + \frac{1}{\sqrt{t}}\right)^k V_{D_2}(t)$$
$$= V_{O_{k+1}}(t)V_{D_2}(t). \quad (11.2.3)$$

Combining (11.2.3) and (11.2.2) gives us

$$V_{K_1 \# K_2}(t) = f_1(t)V_O(t)V_{D_2}(t) + f_2(t)V_{O_2}(t)V_{D_2}(t) + \cdots$$
$$+ f_m(t)V_{O_m}(t)V_{D_2}(t)$$
$$= \{f_1(t)V_O(t) + f_2(t)V_{O_2}(t) + \cdots + f_m(t)V_{O_m}(t)\}V_{D_2}(t)$$
$$= V_{K_1}(t)V_{K_2}(t).$$

The following theorem follows immediately from Theorem 11.2.2, so we leave the proof as an exercise for the reader:

Theorem 11.2.3.

$$V_{D_1 \amalg D_2}(t) = -\left(\sqrt{t} + \frac{1}{\sqrt{t}}\right)V_{D_1}(t)V_{D_2}(t).$$

Theorem 11.2.4.
Suppose $-K$ *is the knot with the reverse orientation to that on* K, *then*

$$V_{-K}(t) = V_K(t).$$

Proof
In order to calculate $V_{-K}(t)$, we may use the same skein tree diagram as for the calculation of $V_K(t)$ (why?). Hence the result follows.

As a consequence of the above theorem, the Jones polynomial is not a useful tool in the study of whether or not a knot is invertible. However, the Jones polynomial is a powerful tool in the study of amphicheirality of a knot.

Theorem 11.2.5.
Suppose K^* *is the mirror image of a knot (or link)* K, *then*

$$V_{K^*}(t) = V_K(t^{-1}).$$

Therefore, if a knot K is amphicheiral, then $V_K(t) = V_K(t^{-1})$, i.e., $V_K(t)$ is symmetric.

<u>Proof</u>

Suppose D is a regular diagram of K and D^* is its mirror image. If the skein tree diagram of D is, say, R, then we can form the skein tree diagram of D^*, R^*, as follows: When we perform a skein operation at a crossing point, c, of D to make R, at the equivalent crossing point of D^* also perform a skein operation, so forming R^*. Since the signs of the crossing point c of D and the equivalent crossing point in D^* are opposite, the coefficients assigned at this juncture to R and R^* differ. If at a certain segment α of R we have assigned t^2 (or t^{-2}), then at the equivalent segment, α^*, of R^* we assign t^{-2} (or t^2). On the other hand, if have assigned tz (or $-t^{-1}z$), then equivalently at α^* we assign $-t^{-1}z$ (or tz).

Since this change

$$t^2 \rightleftharpoons t^{-2} \quad \text{and} \quad tz \rightleftharpoons -t^{-1}z$$

[note:

$$tz = t\left(\sqrt{t} - \frac{1}{\sqrt{t}}\right) \rightleftharpoons -t^{-1}z = -t^{-1}\left(\sqrt{t} - \frac{1}{\sqrt{t}}\right) = \frac{1}{t}\left(\frac{1}{\sqrt{t}} - \sqrt{t}\right)\Big]$$

is nothing but the replacement of t by t^{-1}, it follows that $V_{K^*}(t) = V_K(t^{-1})$.   ∎

Example 11.2.1. The Jones polynomial of the right-hand trefoil knot K is $V_K(t) = t + t^3 - t^4$, and since this is not symmetric by the above theorem, K is not amphicheiral.

Example 11.2.2. The Jones polynomial of the figure 8 knot *is* symmetric (cf. Exercise 11.1.1).

Exercise 11.2.1. Calculate the Jones polynomial of the knot K in Figure 6.4.2 and show it is not symmetric. Therefore, this knot is not amphicheiral; however, $\sigma(K) = 0$ (cf. Exercise 6.4.4).

Caveat lector, a symmetric Jones polynomial does not imply that the knot is amphicheiral.

Exercise 11.2.2. Show that the Jones polynomial of the knot, K, in Figure 11.2.2 is

$$V_K(t) = t^{-3} - t^{-2} + t^{-1} - 1 + t - t^2 + t^3.$$

However, prove, by calculating its signature, that this knot is not amphicheiral.

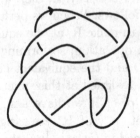

Figure 11.2.2

Example 11.2.3. The Alexander polynomials of the 2-bridge knots K_1 and K_2 of type (15,4) and (15,7), respectively, are the same, namely, $4t^{-1} - 7 + 4t$. But the Jones polynomials of these two knots are different:

$$V_{K_1}(t) = t - 2t^2 + 3t^3 - 2t^4 + 3t^5 - 2t^6 + t^7 - t^8$$
$$V_{K_2}(t) = t - t^2 + 2t^3 - 2t^4 + 2t^5 - 2t^6 + 2t^7 - t^8 + t^9 - t^{10},$$

and so they are not equivalent. (In fact, this also follows from Theorem 9.3.3.)

Example 11.2.4. Calculations of the Jones polynomial have shown that it is non-trivial in the cases when the Alexander polynomial is trivial, i.e., equal to 1. However, whether this is true in general is still a *conjecture*. An example of this phenomenon is the Kinoshita-Terasaka knot (Figure 6.3.2); its Jones polynomial is

$$V_K(t) = -t^{-4} + 2t^{-3} - 2t^{-2} + 2t^{-1} + t^2 - 2t^3 + 2t^4 - 2t^5 + t^6.$$

More generally, at present there are no non-trivial examples of knots that have the Jones polynomial equal to 1. If, in fact, we can prove that such an example does not exist, then the Jones polynomial, at the very least and not insignificantly, will enable us to determine whether or not a knot is trivial. In this regard, we offer the following exercise for the ambitious reader:

Exercise (no known solution). Does there exist a non-trivial knot K whose Jones polynomial is equal to 1? Further, is there a non-trivial μ-component link, L, such that

$$V_L(t) = (-1)^{\mu-1}\left(\sqrt{t} + \frac{1}{\sqrt{t}}\right)^{\mu-1}?$$

Although the Jones polynomial is a strong invariant, it is not a complete invariant. That is to say, there exist an infinite number of non-equivalent knots that have the same Jones polynomial.

Example 11.2.5 [Kan].   The Jones polynomial of the two knots K_1 and K_2 shown in Figure 11.2.3(a) and (b), respectively, are the same, namely, are equal to $(t^{-2} - t^{-1} + 1 - t + t^2)^2$. However, their Alexander polynomials are

$$\Delta_{K_1}(t) = (t^{-1} - 3 + t)^2$$
$$\Delta_{K_2}(t) = -t^{-3} + 3t^{-2} - 5t^{-1} + 7 - 5t + 3t^2 - t^3.$$

Therefore, K_1 and K_2 are not equivalent.

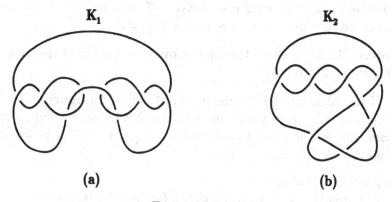

Figure 11.2.3

This example shows that even though there are cases when knots can be distinguished by the Jones polynomial and not by the Alexander polynomial, the reverse is also true. So asking which is the more powerful invariant is to a certain degree a meaningless question.

Example 11.2.6 [LM3].   The two knots K_1 and K_2 in Figure 11.2.4(a) and (b) are not equivalent; however, their Jones polynomial are the same,

$$V_K(t) = -t^{-3} + 2t^{-2} - 3t^{-1} + 5 - 4t + 4t^2 - 3t^3 + 2t^4 - t^5,$$

and their Alexander polynomials are also equal,

$$\Delta_K(t) = 2t^{-2} - 6t^{-1} + 9 - 6t + 2t^2.$$

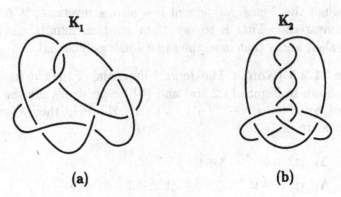

(a)   (b)

Figure 11.2.4

Showing they are not equivalent can be done by calculating their respective unknotting numbers, $u(K_1) = 2$ and $u(K_2) = 1$. However, it is not an easy matter to show that $u(K_1) = 2$.

Exercise 11.2.3. Show that the unknotting number of the knot K_2 in Figure 11.2.4(b) is 1.

If we substitute for t some numerical value, for example, ω, the pth root of unity, then the Jones polynomial is in fact related to other known knot invariants, at least for relatively small values of p. We shall discuss some of the known cases.

Proposition 11.2.6.
   Suppose that L is a μ-component (oriented) link, then

$$V_L(1) = (-2)^{\mu-1}.$$

A consequence of this proposition is that the Jones polynomial can never be zero.

Proof
   Since if we substitute $t = 1$ into the skein formula (11.1.1) we obtain $V_{L_+}(1) - V_{L_-}(1) = 0$, it follows that

$$V_{L_+}(1) = V_{L_-}(1) = V_{O_\mu}(1)$$

(cf. the proof of Proposition 6.3.1). However, due to Proposition 11.1.1,

$$V_{O_\mu}(1) = (-2)^{\mu-1} = V_L(1).$$

Proposition 11.2.7.

If K is a knot or link, then

$$V_K(-1) = (-1)^{\mu(K)-1}\Delta_K(-1),$$

where $\mu(K)$ denotes the number of components of K.

<u>Proof</u>

If we let $t = -1$ in (11.1.1), then

$$-V_{K_+}(-1) + V_{K_-}(-1) = \left(\sqrt{-1} - \frac{1}{\sqrt{-1}}\right)V_{K_0}(-1). \qquad (11.2.4)$$

On the other hand, using Theorem 6.2.1, we obtain that

$$\Delta_{K_+}(t) - \Delta_{K_-}(t) = \left(\sqrt{t} - \frac{1}{\sqrt{t}}\right)\Delta_{K_0}(t). \qquad (11.2.5)$$

Substituting $t = -1$ in (11.2.5), we obtain

$$\Delta_{K_+}(-1) - \Delta_{K_-}(-1) = \left(\sqrt{-1} - \frac{1}{\sqrt{-1}}\right)\Delta_{K_0}(-1). \qquad (11.2.6)$$

Next, we multiply (11.2.6) by the factor $-(-1)^{\mu(K_+)-1}$, and hence

$$-(-1)^{\mu(K_+)-1}\Delta_{K_+}(-1) + (-1)^{\mu(K_+)-1}\Delta_{K_-}(-1)$$

$$= -(-1)^{\mu(K_+)-1}\left(\sqrt{-1} - \frac{1}{\sqrt{-1}}\right)\Delta_{K_0}(-1).$$

Since $\mu(K_-) = \mu(K_+)$ and $\mu(K_0) = \mu(K_+) + 1$ or $\mu(K_+) - 1$,

$$-(-1)^{\mu(K_+)-1}\Delta_{K_0}(-1) = (-1)^{\mu(K_0)-1}\Delta_{K_0}(-1)$$

and

$$-(-1)^{\mu(K_+)-1}\Delta_{K_+}(-1) + (-1)^{\mu(K_-)-1}\Delta_{K_-}(-1)$$

$$= (-1)^{\mu(K_0)-1}\left(\sqrt{-1} - \frac{1}{\sqrt{-1}}\right)\Delta_{K_0}(-1).$$

$$(11.2.7)$$

Since (11.2.4) and (11.2.7) are exactly the same skein relation, the required result follows.   ∎

A few more cases are given in the next proposition, which we will not prove.

Proposition 11.2.8.

Suppose K is a knot. Then

(1) [J1]   If ω is a primitive 3th root of unity, then $V_K(\omega) = 1$;

(2) [Muk] Let $i = \sqrt{-1}$ (i.e., a primitive 4th root of unity),

        (a) if $\Delta_K(-1)$ is of the form $8k \pm 1$, $V_K(i) = 1$;

        (b) if $\Delta_K(-1)$ is of the form $8k \pm 3$, $V_K(i) = -1$.

Note in the above $\Delta_K(-1)$ is always an odd integer. If K is a link, then we can obtain similar formulae.

Also, if ξ is a primitive 6th root of unity, then we may calculate $V_K(\xi)$ [LM1]. A general formula for an arbitrary p^{th} root of unity has yet to be discovered, this task seems to be by no means an easy one.

The Alexander polynomial of a link L is an invariant of an oriented link. If, however, we change the orientation of a single component of L, then the polynomial is completely altered. This change is quite difficult to categorize. On the other hand, the Jones polynomial of L is an invariant that is almost unrelated to the assigned orientation, because if we change the orientation of a component, we can exactly describe the effect on the Jones polynomial.

Theorem 11.2.9 [LM2].

Suppose $L = \{K_1, K_2, \ldots, K_\mu\}$ is a μ-component (oriented) link. Further, suppose $\widehat{L} = \{K_1, K_2, \ldots, K_{\mu-1}, -K_\mu\}$ is the link with the same orientation as L, except the orientation on the component K_μ has been reversed. Then

$$V_{\widehat{L}}(t) = t^{-3l}V_L(t),$$

where $l = \sum_{i=1}^{\mu-1} \text{lk}(K_\mu, K_i)$.

Exercise 11.2.4. Suppose L and L$'$ are two oriented links obtained from two different orientations on the torus link $K_{4,2}$. Calculate the Jones polynomial of each, and so check the result of Theorem 11.2.9. Further, calculate their Alexander polynomials and then compare the differences between the two cases.

Exercise 11.2.5. Reread the proofs of Theorems 11.2.2, 11.2.4, and 11.2.5, and along the same lines reprove Theorems 6.3.3 and 6.3.5.

§3 The skein invariants

If we look carefully at the skein relation (11.1.1) that defines the Jones polynomial and compare it in particular with (6.2.1), then it is possible to perceive that the coefficient of $V_{K_+}(t)$ and $V_{K_-}(t)$ need not necessarily be limited to $\frac{1}{t}$ and t. In general, it seems that we may take an arbitrary function of t and allow it to vary. In other words, it should be possible to choose a coefficient with more than one indeterminate. Actually, in this way it is possible to define the most general polynomial.

Definition 11.3.1.   Suppose K is a (oriented) knot (or link) and D is a regular diagram of K. Then we may, by means of the following axioms, define a polynomial of K, $S_K(x, y, w)$, in the three indeterminates x,y,w that may have negative exponents.

(1) If K is the trivial knot O, then $S_O(x, y, w) = 1$.

(2) With regard to the skein diagrams D_+, D_-, D_0,

the following equality holds: (11.3.1)

$$x S_{D_+}(x, y, w) - y S_{D_-}(x, y, w) = w S_{D_0}(x, y, w).$$

The polynomial $S_K(x, y, w)$ is an invariant of K.

Before we calculate $S_K(x, y, w)$ for several knots, let us consider a slight simplification of S_K. From the above definition it seems that S_K is a polynomial in 3 indeterminates; however, essentially it is a polynomial in only 2 indeterminates. In order to show this, let us consider the 2-variable polynomial $P_K(v, z)$ defined by (11.3.2).

(1) For the trivial knot O, $P_O(v, z) = 1$

(2) With regard to the skein diagrams D_+, D_-, D_0,

the following formula holds: (11.3.2)

$$\tfrac{1}{v} P_{D_+}(v, z) - v P_{D_-}(v, z) = z P_{D_0}(v, z).$$

That $P_K(v, z)$ and $S_K(x, y, w)$ are intrinsically the same follows from the two equalities in (11.3.3).

Exercise 11.3.1.   Show that

$$(1)\ S_K\left(\frac{1}{v}, v, z\right) = P_K(v, z)$$

$$(2)\ P_K\left(\frac{\sqrt{y}}{\sqrt{x}}, \frac{w}{\sqrt{x}\sqrt{y}}\right) = S_K(x, y, w).$$

(11.3.3)

In general, a polynomial defined by skein relations is called a *skein polynomial*. $P_K(v, z)$ is the most generalized form of a skein polynomial and is usually called the HOMFLY polynomial; the initials stand for the surnames of the mathematicians who, at roughly the same time, discovered this polynomial.

If we compare the skein relation definitions of the Alexander and the Jones polynomial [Definitions (6.2.1) and (11.1.1)] with the skein relation of $P_K(v, z)$, (11.3.2), then the following is an easy consequence:

Proposition 11.3.1.

Suppose that K is a (oriented) knot (or link), then

$$(1)\ V_K(t) = P_K\!\left(t, \sqrt{t} - \frac{1}{\sqrt{t}}\right)$$

$$(2)\ \Delta_K(t) = P_K\!\left(1, \sqrt{t} - \frac{1}{\sqrt{t}}\right)$$

So it may be said that $V_K(t)$ and $\Delta_K(t)$ are special cases of $P_K(v, z)$. However, $\Delta_K(t)$ is not a special case of $V_K(t)$. As we have already mentioned, they are essentially different polynomials.

To calculate $S_K(x, y, w)$ and $P_K(v, z)$ for an arbitrary knot (or link) K, it is better to use the skein tree diagram.

Exercise 11.3.2. Show
$$P_{OO}(v, z) = \frac{1 - v^2}{vz}.$$

Exercise 11.3.3. Show, using a skein tree diagram, that the skein polynomial of the right-hand trefoil knot K is

$$P_K(v, z) = 2v^2 - v^4 + v^2 z^2.$$

§4 The Kauffman polynomial

We know that if we can transform one knot diagram to another knot diagram via the Reidemeister moves $\Omega_1, \Omega_2, \Omega_3$ or their inverses, then they are equivalent. So a possible approach to show that the function we have developed is a knot invariant is to show it remains unchanged under these Reidemeister moves. Hence, we must investigate in what way all *three* Reidemeister moves affect the function, especially since their characteristics differ completely in several aspects. For example, we know that if we apply Ω_3 or Ω_3^{-1}, then the number of crossing points of the regular diagram D remains unchanged. A further important consideration is that even if we give D an orientation, the Tait number (Definition 4.5.2), w(D), remains unchanged when we apply either Ω_2 or Ω_3, or its inverse. However, the Tait number itself is not a knot invariant; therefore, it would seem it is not sufficient to show two regular diagrams are invariant under just Ω_2 and Ω_3 and their inverses. The question is, How "far" are we from a knot invariant if we restrict ourselves to just to Ω_2 or Ω_3 or its inverse? In this regard, L. Kauffman, with extreme perspicacity, arrived at the following important observation:

Definition 11.4.1. Let us call the Reidemeister moves Ω_2 or Ω_3 and their inverses *regular moves*. Then, if we can obtain a regular diagram D' by applying these regular moves a finite number of times to a regular diagram D of some knot (or link), we say D and D' are *regular equivalent*.

Kauffman's principle.

Suppose a function, f, with indeterminate t (for a multivariable function, see Definition 11.4.2) is invariant under the regular moves. If we choose m suitably (it will depend on the regular diagram), then $t^m f$ is an invariant of knots (and links).

Exercise 11.4.1. Show that the Tait number (or the writhe) of an oriented regular diagram is invariant under regular equivalence.

Let us explain the above principle by considering a couple of examples. As in the first example, we shall consider the Kauffman bracket polynomial defined below. It is essentially the same as the Jones polynomial, however, this polynomial has certain special properties for some particular types of knots, such as alternating knots and links. Consequently, it has had a significant impact on the study of alternating knots (and links); we shall discuss this in more detail in the next section.

Suppose K is an *unoriented* knot (or link) and D is a regular diagram of K. Cut (splice) each crossing point of D in the two ways shown in Figure 11.4.1 (*nota bene*, this splicing process is independent of the sign of the crossing point).

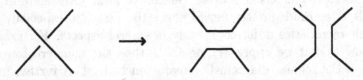

Figure 11.4.1

The reason we need to make sure that our knot (or link) is unoriented is because if we assign an orientation to D and then cut, the orientation on the new regular diagram will not longer be compatible, see Figure 11.4.2.

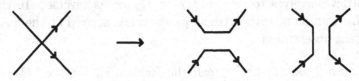

Figure 11.4.2

We shall now use the above process of splicing a crossing point to define the Kauffman bracket polynomial.

Theorem 11.4.1 [Kau1].

Let D be an unoriented regular diagram of a knot or link K. Then there exists a unique one-variable integer polynomial $P_D(A)$ (with possibly negative exponents) that satisfies the following four conditions:

    (1) *$P_D(A)$ is invariant under regular equivalence.*

    (2) *If D is the trivial diagram O of a trivial knot, then*

$$P_O(A) = 1. \tag{11.4.1}$$

    (3) *If D consists of two split regular diagrams D_1, D_2, i.e., $D = D_1 \amalg D_2$, then*

$$P_D(A) = -(A^2 + A^{-2})P_{D_1}(A)P_{D_2}(A). \tag{11.4.2}$$

    (4) *Let D, $\widehat{D}$, $\widehat{D}'$ be the skein diagrams given in Figure 11.4.3. Then the following equality holds:*

$$P_D(A) = AP_{\widehat{D}}(A) + A^{-1}P_{\widehat{D}'}(A). \tag{11.4.3}$$

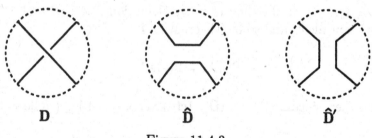

Figure 11.4.3

$P_D(A)$ is called *Kauffman's bracket polynomial*, as noted from the theorem defined on the *regular diagram* D of a knot or link. For example, (11.4.1) does not mean that $P_K(A) = 1$ for the trivial knot K. In fact, to evaluate $P_D(A)$ for $D = \bigcirc\bigcirc$, we must use (11.4.3) to eventually obtain $P_{\bigcirc\bigcirc}(A) = -A^{-3}$. Therefore, $P_D(A)$ is not invariant under the first Reidemeister move, Ω_1. However, it is possible to define an invariant from $P_D(A)$ that is also invariant under Ω_1; this is an implication of Kauffman's principle.

Theorem 11.4.2.

Suppose D is an oriented regular diagram of an oriented knot (or link) K. If $P_D(A)$ is the Kauffman bracket polynomial of the "unoriented" diagram D, and $w(D)$ is the Tait number (writhe) of D, then define

$$\widehat{P}_D(A) = (-A^{-3})^{w(D)} P_D(A). \qquad (11.4.4)$$

Then $\widehat{P}_D(A)$ is an invariant of an oriented knot (or link), denoted by $\widehat{P}_K(A)$.

If we substitute $A = t^{-\frac{1}{4}}$, then $\widehat{P}_K(A)$ coincides with the Jones polynomial $V_K(t)$ of K. Namely,

$$\widehat{P}_K(t^{-\frac{1}{4}}) = V_K(t). \qquad (11.4.5)$$

Therefore, $\widehat{P}_K(A)$ is essentially the same as the Jones polynomial. [We should note that $P_D(A)$ is multiplied by $(-A^{-3})^{w(D)}$ to eliminate the effect on $P_D(A)$ of the kink, since $P_{\bigcirc\bigcirc}(A) = -A^{-3}$.]

<u>Proof of Theorem 11.4.2.</u>

Suppose that D' is a regular diagram of K that has been obtained by performing a single Reidemeister move on D. Then it is sufficient to show

$$\widehat{P}_D(A) = \widehat{P}_{D'}(A).$$

Firstly, let us suppose that D′ has been obtained from D by performing Ω_2, Ω_3 or their inverses. By Definition 11.4.1, D and D′ are regular equivalent, and so by Theorem 11.4.1,

$$P_D(A) = P_{D'}(A).$$

Further, since $w(D) = w(D')$ (cf. Exercise 11.4.1), it follows that

$$\widehat{P}_D(A) = \widehat{P}_{D'}(A).$$

This leaves the case of D′ obtained by applying Ω_1 (or Ω_1^{-1}) to D.

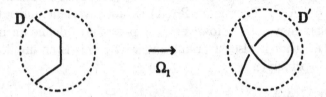

Figure 11.4.4

Since D′ has an extra crossing point, to evaluate $P_{D'}(A)$ we need to use the following skein tree diagram.

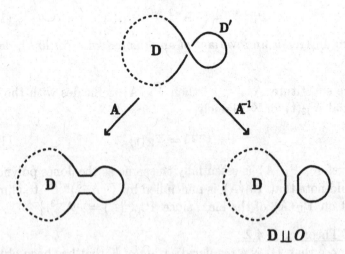

Then, by (11.4.2) and (11.4.3), we have

$$P_{D'}(A) = AP_D(A) + A^{-1}P_D(A)(-(A^2 + A^{-2})) = -A^{-3}P_D(A).$$

Irrespective of how we assign the orientation to D, the sign of the new crossing point is -1. Therefore, $w(D') = w(D) - 1$. This fact, in conjunction with (11.4.4), allows us to write the following:

$$\widehat{P}_{D'}(A) = (-A^{-3})^{w(D')}P_{D'}(A) = (-A^{-3})^{w(D)-1}(-A^{-3})P_D(A)$$

$$= (-A^{-3})^{w(D)}P_D(A) = \widehat{P}_D(A).$$

$\blacksquare$

Exercise 11.4.2. Check that $\widehat{P}_K(t^{-\frac{1}{4}}) = (-t^{\frac{3}{4}})^{w(D)}P_D(t^{-\frac{1}{4}})$ satisfies Axiom 2 in Definition 11.1.1 and show that $\widehat{P}_D(t^{-\frac{1}{4}}) = V_K(t)$.

Example 11.4.1. To evaluate $P_D(A)$, $\widehat{P}_D(A)$ for a regular diagram of the positive Hopf link, L, we shall use the skein tree diagram, but first we should note that $w(D) = 2$.

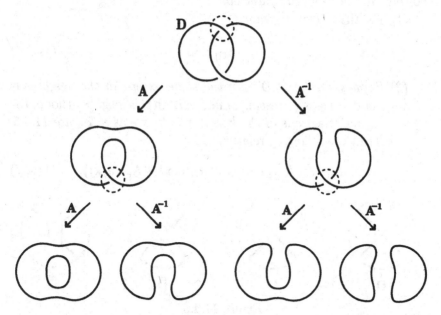

Therefore, by reading off the coefficients from the skein tree diagram,

$$P_D(A) = A^2(-(A^2 + A^{-2})) + 1 + 1 + A^{-2}(-(A^2 + A^{-2}))$$

$$= -A^4 - A^{-4}.$$

Since $w(D) = 2$, $\widehat{P}_D(A) = (-A^{-3})^2(-A^4 - A^{-4}) = -A^{-2} - A^{-10}$, and so $\widehat{P}_D(t^{-\frac{1}{4}}) = -t^{\frac{1}{2}} - t^{\frac{5}{2}}$, which not surprisingly is the same as the Jones polynomial of L.

Exercise 11.4.3. Evaluate $P_D(A)$ for the regular diagram of the right-hand trefoil knot K, Figure 0.1(c). Subsequently evaluate $\widehat{P}_D(A)$ to check that $\widehat{P}_D(t^{-\frac{1}{4}}) = V_K(t)$.

As an extension of the Kauffman bracket polynomial (and the Jones polynomial) of a knot (or link) K, it is possible to define a two-variable integer polynomial (with possible negative exponents), $F_K(a, x)$, using similar arguments as the above. This will form our second example of the efficacy of Kauffman's principle.

Theorem 11.4.3 [Kau2].

Suppose D is a regular diagram of a knot (or link) on which no orientation has been assigned. Then there exists the 2-variable polynomial $\Lambda_D(a, x)$ that is invariant under the regular moves, provided the following 3 conditions are satisfied.

(1) For the trivial diagram O,

$$\Lambda_O(a, x) = 1. \tag{11.4.6}$$

(2) Suppose $D, D', \widehat{D}, \widehat{D}'$ are the same except in the neighbourhood of a single crossing point. Within this neighbourhood, the regular diagrams of the knot (or link) are as in Figure 11.4.5. Then the following relation holds:

$$\Lambda_D(a, x) + \Lambda_{D'}(a, x) = x\{\Lambda_{\widehat{D}}(a, x) + \Lambda_{\widehat{D}'}(a, x)\}. \tag{11.4.7}$$

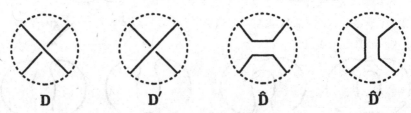

$$D \qquad\qquad D' \qquad\qquad \widehat{D} \qquad\qquad \widehat{D}'$$

Figure 11.4.5

(3) Suppose $D, \overline{D}, D_0$ are the same except within the neighbourhood of a single crossing point. Within this neighbourhood the regular diagrams of the knot (or link) are as in Figure 11.4.6. Then the following formulae hold:

$$(i)\ \Lambda_D(a, x) = a\Lambda_{D_0}(a, x)$$

$$(ii)\ \Lambda_{\overline{D}}(a, x) = a^{-1}\Lambda_{D_0}(a, x). \tag{11.4.8}$$

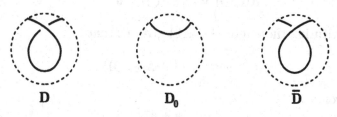

$$D \qquad\qquad D_0 \qquad\qquad \bar{D}$$

Figure 11.4.6

This 2-variable polynomial $\Lambda_D(a, x)$ is also unique. As in the case of the Kauffman bracket polynomial, this invariant is defined from the regular diagram. So, as before we do not define $\Lambda_K(a, x) = 1$ if K is the trivial knot. In fact, $\Lambda_D(a, x)$ is not invariant under the first Reidemeister move, Ω_1. Therefore, in order for $\Lambda_D(a, x)$ to be also invariant under Ω_1, we need to define a new polynomial.

Definition 11.4.2. Suppose K is an *oriented* knot (or link) and D is a (oriented) regular diagram for K. Further suppose that $w(D)$ is the Tait number of D. Then, let

$$F_D(a, x) = a^{w(D)} \Lambda_D(a, x). \qquad (11.4.9)$$

Theorem 11.4.4.
$F_D(a, x)$ *is an invariant of the oriented knot (or link) K, independent of its regular diagram D. This invariant is called the (2-variable) Kauffman polynomial.*

The proof of this theorem runs along the same lines as the proof of Theorem 11.4.2, so it is left as an exercise for the reader.

Let us now give an example of a calculation of the Kauffman polynomial. For the sake of clarity, we shall write $\Lambda_D(a, x)$ as $\Lambda(D)$.

Example 11.4.2. Suppose $L = \{OO\}$. In order to calculate $\Lambda(D)$ we need to use (11.4.7),

$$\Lambda(\infty) + \Lambda(\infty) = x\{\Lambda(\subset\supset) + \Lambda(OO)\}, \qquad (11.4.10)$$

where ∞ is a regular diagram of the trivial knot, but it is not regular equivalent to O, so $\Lambda(\infty) \neq \Lambda(O)$.

By (11.4.8),

$$\Lambda(\infty) = a\Lambda(O) = a$$

and

$$\Lambda(\text{CO}) = a^{-1}\Lambda(\text{O}) = a^{-1}.$$

Substituting these into (11.4.10) gives us that

$$a^{-1} + a = x\{1 + \Lambda(\text{OO})\}.$$

Therefore,

$$\Lambda(\text{OO}) = \frac{a + a^{-1}}{x} - 1.$$

On the other hand, $w(D) = 0$, so the Kauffman polynomial of L is

$$F_L(a, x) = \Lambda_L(a, x) = x^{-1}(a + a^{-1}) - 1.$$

Exercise 11.4.4. Show that the Kauffman polynomial of the right-hand trefoil knot K is

$$F_K(a, x) = a^{-3}\{x^2(a + a^{-1}) + x(1 + a^{-2}) - (2a + a^{-1})\}$$

[Hint: Since OD and OO are regular equivalent, $\Lambda(\text{OD}) = \Lambda(\text{OO})$.]

As in the case of the bracket polynomial, the Kauffman polynomial and Jones polynomial are related.

Theorem 11.4.5 [Kau2].
Suppose K is a (oriented) knot (or link), then

$$F_K(-t^{-\frac{3}{4}}, t^{\frac{1}{4}} + t^{-\frac{1}{4}}) = V_K(t).$$

Exercise 11.4.5. Check the validity of Theorem 11.4.5 by substituting the above values into the Kauffman polynomial calculated in Exercise 11.4.4.

Hence, by the above theorem the Kauffman polynomial, as well as the Kauffman bracket polynomial, may be thought of as extensions of the Jones polynomial. As in the case of the Jones polynomial, the two Kauffman polynomials are not complete invariants, because there exist many non-equivalent knots with the same Kauffman polynomials and Kauffman bracket polynomials.

In general, no matter how we construct the skein polynomials, there will always exist a knot (or link) that cannot be distinguished by the given polynomial. Therefore, it is impossible to find a skein polynomial that will be a complete knot invariant.

The Kauffman principle, as we shall see in the next chapter, plays an important role when a knot invariant is defined from an exactly solvable model of the type found in statistical mechanics.

§5 The skein polynomials and classical knot invariants.
(Alternating knots and the Tait conjectures)

Although in definition an alternating knot is a simple matter, from the inchoate stages of knot theory to the present, this knot has been thought to be *non-trivial*. Furthermore, a reduced alternating diagram can readily be understood intuitively, i.e., it is a regular diagram that we cannot change into one with a fewer number of crossings. However, it is only quite recently that this has been proven to the standards of mathematical rigour. The study of alternating knots has a long history, and over this period a panoply of its characteristics had been proven, but the Tait conjectures had haunted and proved unyielding.

The new invariants, the Jones polynomial and the skein polynomials, which have been the focus of our attention, in this chapter, in actual fact have played the crucial rôle in solving the primary Local problems. These were the problems/conjectures that concerned Tait in the 19th century and from which germinated modern knot theory. The conjectures deal with a specific type of knot, namely, the alternating knots. In order to develop these conjectures, we shall in this chapter take a more detailed look at Local problems for alternating knots and links.

The *alternating* in alternating knot signifies a geometric characteristic, i.e., the alternating nature of the crossings. It is the case that on occasions geometric invariants of knots can be determined by algebraic invariants. The next theorem is a classic example of such a case.

Theorem 11.5.1 [Mus1].

Suppose K is an alternating knot. The genus of K, $g(K)$, is equal to the degree of the Alexander polynomial of K. Further, the Seifert surface constructed from an alternating diagram has the minimal genus $g(K)$.

This theorem shows that the *genus* of K, a geometric invariant, is completely determined by the (maximum) degree of the Alexander polynomial, an algebraic invariant. In the case of torus knots, we know (Theorem 7.5.2) that a similar result holds. This gives a further distinct example of a geometric characteristic, namely, a torus knot, being reflected in its algebraic nature.

In addition, the Alexander polynomial for an alternating knot has a special form.

Theorem 11.5.2 [Mus2].

Suppose K is an alternating knot and

$$\Delta_K(t) = a_{-m}t^{-m} + a_{-m+1}t^{-m+1} + \ldots + a_m t^m$$

is its Alexander polynomial $(a_m \neq 0 \neq a_{-m})$. *Then*

(1) $a_{-m}, a_{-m+1}, \ldots, a_m$ *are never equal to zero;*

(2) *the sign of two consecutive coefficients alternates, i.e.,*

$$a_i a_{i+1} < 0 \quad (i = -m, -m+1, \ldots, m-1).$$

Exercise 11.5.1. By looking again at the calculations of the trefoil knot and the figure 8 knot, show the veracity of the above theorem.

Exercise 11.5.2. If $K(q, r)$ is a torus knot with $\min\{|q|, |r|\} \geq 3$, show that it cannot be an alternating knot.

From condition (2) in the above, the Alexander polynomial of an alternating knot is an *alternating* polynomial.[14]

The signature of an alternating knot is also comparatively easy to calculate. However, since a few preliminary preparations are required, we shall not discuss the details here but refer the reader to Murasugi [Mus3].

Also, by looking at the reduced regular diagram, it is relatively easy to determine whether or not a given alternating knot (or link) is prime.

Theorem 11.5.3 [Men].

Suppose D is a reduced alternating diagram of an alternating knot (or link) K. If K is not prime, then there exists a circle C on the plane that intersects D in exactly two points, and this circle C divides D into two non-trivial (1, 1)-tangles (one of the tangles lies within the circle, while the lies without).

Caveat lector, Theorem 11.5.3 does not hold if D is not an alternating diagram.

Exercise 11.5.3. Show that 2-bridge knots (or links) are prime knots.

Exercise 11.5.4. Determine whether the two knots in Figure 11.5.1 are prime knots.

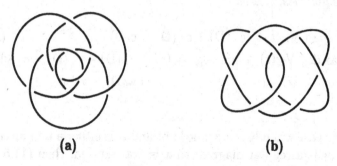

(a)   **(b)**

Figure 11.5.1

If we take a second look at Theorem 11.5.1, this theorem seems to suggest that perhaps the *degree* of a polynomial is the most important aspect of the polynomial as an invariant. Previously we showed that the degree of the Alexander polynomial of a knot K is not greater than its genus (Theorem 6.3.7). The natural question is, In what way (if any) is the degree of the Jones polynomial of K an invariant of K or on what knot theoretical property does it depend? Since the Jones polynomial may also have negative exponents in addition to defining its maximum degree, $\max \deg V_K(t)$, we also define the minimum degree, $\min \deg V_K(t)$. We can prove that these can be estimated using other invariants.

So, suppose D is a regular diagram of an oriented knot (or link) K. Further, let $c_+(D)$ and $c_-(D)$ be the number, respectively, of positive and negative crossing points of D. Finally, let $\sigma(K)$ be the signature of K.

Theorem 11.5.4 [Mus5].

For an arbitrary connected regular diagram D of K, the following inequality holds:

$$\min \deg V_K(t) \geq -c_-(D) - \frac{1}{2}\sigma(K) \tag{11.5.1}$$

$$\max \deg V_K(t) \leq c_+(D) - \frac{1}{2}\sigma(K). \tag{11.5.2}$$

Therefore, if we set $\operatorname{span} V_K(t) = \max \deg V_K(t) - \min \deg V_K(t)$, *then the following holds:*

$$\operatorname{span} V_K(t) \leq c_+(D) + c_-(D) = c(D). \tag{11.5.3}$$

If K is a (non-split) alternating knot (or link) and D is a reduced alternating diagram for K, then in (11.5.1) and (11.5.2) the inequalities become equalities; therefore,

$$\text{span } V_K(t) = c_+(D) + c_-(D) = c(D) \qquad (11.5.4)$$
$$\max \deg V_K(t) + \min \deg V_K(t) = c_+(D) - c_-(D) - \sigma(K)$$
$$= w(D) - \sigma(K). \qquad (11.5.5)$$

In the case when K is a prime knot and D is either a non-alternating diagram or is a regular diagram that is not reduced, then (11.5.4) and (11.5.5) do not hold.

Exercise 11.5.5. Confirm (11.5.3) and (11.5.4) hold for all the knots listed in Appendix (II).

Exercise 11.5.6. Let L be the oriented 2-bridge link of type (16,5) and L′ the oriented 2-bridge link obtained from L by reversing the orientation of one of the components. [L′ is of type (16, −11).] Determine the maximal and minimal degrees of the Jones polynomials of L and L′, and confirm these values are consistent with Theorem 11.2.9.

An application of Theorem 11.5.4 is in the proof of the next two "primary" theorems, called Tait's first and second conjectures, respectively.

Theorem 11.5.5 (Tait's First Conjecture).
A reduced alternating diagram is the minimum diagram of its alternating knot (or link). Moreover, the minimum diagram of a prime alternating knot (or link) can only be an alternating diagram. In other words, a non-alternating diagram can never be the minimum diagram of a prime alternating knot (or link).

Proof
If we restrict ourselves to reduced (connected) alternating diagrams, then since (11.5.4) and (11.5.5) always hold, the number of crossing points of D, $c(D) = c_+(D) + c_-(D)$, is fixed. Also span $V_K(t)$ is an invariant of K that is independent of D. Therefore, $c(D)$ in this case is also an invariant of K. Further, by (11.5.3) the number of crossing points cannot decrease below span $V_K(t)$. Hence, D must be a minimum diagram. Now, (11.5.4) can never hold for a non-alternating diagram of a prime alternating knot. So, such a non-alternating diagram can never be a minimum diagram. ∎

In Chapter 4, Section 5 we defined (cf. Definition 4.5.2) the Tait number (or writhe), $w(D)$, with respect to a regular diagram D of a knot (or link) K. This number in general is not an invariant of K. However, if D is a reduced alternating diagram (and hence a minimum diagram), then it becomes an invariant of K. This is the essence of Tait's second conjecture.

Theorem 11.5.6   (Tait's Second Conjecture).

Suppose that D_1 *and* D_2 *are two reduced alternating diagrams of an alternating knot (or link)* K, *then* $w(D_1) = w(D_2)$.

<u>Proof</u>

Let us consider (11.5.5). Since we restrict ourselves to diagrams D that are reduced (connected) alternating diagrams, (11.5.5) holds. Further, the maximum and minimum degrees of $V_K(t)$ and the signature $\sigma(K)$ are invariants of K that are independent of D. Therefore, $w(D)$ is an invariant of K.

∎

Let us now give a straightforward application of this theorem. Suppose D is a reduced alternating diagram of an alternating knot K. Then we can form D^*, the reduced alternating diagram of the mirror image K^* of K, by switching the over and under segments at each crossing point of D. Due to this switch, it is easy to see that the sign of each crossing point of D^* is exactly opposite to the corresponding one on D. Therefore, $w(D^*) = -w(D)$. If, now, we suppose that K is amphicheiral, then since $K \cong K^*$, it is also true that $D \cong D^*$. Hence, $w(D) = w(D^*) = -w(D)$, which implies $w(D) = 0$. Hence, the number of crossing points of D must be even. The thrust of the discussion is encapsulated in the next theorem.

Theorem 11.5.7.

An alternating knot whose minimum number of crossing points is odd can never be amphicheiral.

As yet, an amphicheiral non-alternating knot whose minimum number of crossing points is odd has not been found. So, we cannot say whether or not Theorem 11.5.7 holds for a general non-alternating knot.

The Tait conjectures were initially bruited about by Tait at the beginning of the 20th century after studying copious alternating knots. Tait himself tried to prove them, but without much success.

Tait also put forward another conjecture, which we may call the

Third Tait Conjecture.

Suppose an alternating diagram includes a $(2, 2)$-tangle, as shown in Figure 11.5.2(a).

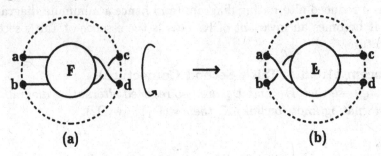

(a)   **(b)**

Figure 11.5.2

Let us fix the four points a,b,c,d, and then rotate this tangle by a half-revolution, Figure 11.5.2(b). Hence, a twist on the right in Figure 11:5.2(a) has moved to the left. Such an operation is called a (Conway) *flype*.

Tait's Third Conjecture.

Suppose D_1 and D_2 are two reduced alternating diagrams of an alternating knot K. Then we can change D_1 into D_2 by performing a finite number of flypes.

This conjecture has very recently shown to be true [MenT].

Exercise 11.5.7. Show that this conjecture implies the second conjecture.

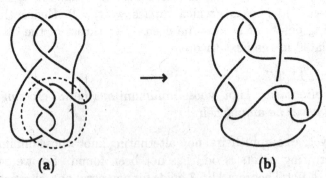

(a)   **(b)**

Figure 11.5.3

Example 11.5.1. Figures 11.5.3(a) and (b) are equivalent regular diagrams of an alternating knot. By performing a flype within the dotted circle in Figure 11.5.3(a), we may transform it into Figure 11.5.3(b).

As was shown above, the degree of the Jones polynomial is clearly related to the number of crossing points of a regular diagram D of a knot (or link) K. A natural question to ask is, What types of invariants of K are the degree of the skein polynomial $P_K(v, z)$ related to? Since $P_K(v, z)$ has two variables, we can define two degrees, one that depends on v, the v-deg $P_K(v, z)$, and the other that depends on z, the z-deg $P_K(v, z)$. In general,

$$\text{z-deg } P_K(1, z) \leq \text{z-deg } P_K(v, z). \tag{11.5.6}$$

Since $P_K(1, z)$ is the Alexander-Conway polynomial [Proposition 11.3.1(2)], its degree is equal to twice the degree of the Alexander polynomial $\Delta_K(t)$. Even if the equality does not hold in (11.5.6), z-deg $P_K(v, z)$ is approximately equal to twice the degree of $\Delta_K(t)$. Therefore, it is v-deg $P_K(v, z)$ that piques our interest.

Proposition 11.5.8 [Mo].
Suppose K is a (oriented) knot (or link) that has been formed from an n-braid β. Then

$$\text{v-span } P_K(v, z) = \max \text{ v-deg } P_K(v, z) - \min \text{ v-deg } P_K(v, z)$$
$$\leq 2(n - 1). \tag{11.5.7}$$

In general, the equality does not hold in (11.5.7). However, there are quite a few knots for which the equality does hold. For such a knot K, since K can never be represented by a braid that has fewer braid strings than n, the braid index of K is exactly n. For example, for 2-bridge knots their braid index is completely determined by the degree of the skein polynomial.

Theorem 11.5.9 [Mus6].
The braid index, b(K), of an oriented 2-bridge knot (or link) K is

$$\frac{1}{2}\{\text{v-span } P_K(v, z)\} + 1.$$

However, excluding these 2-bridge knots, it is not known for which other knots (if any) their braid index is completely determined by $P_K(v, z)$.

Exercise 11.5.8. Evaluate $P_K(v, z)$ for the figure 8 knot K and confirm Theorem 11.5.9 for K. (Hint: K is a 2-bridge knot with braid index 3.)

Knots via Statistical Mechanics

The motivation behind statistical mechanics is to try to understand, by using statistical methods, macroscopic properties – the easiest example being to determine what happens to water in a kettle when we boil it – by looking at the microscopic properties, i.e., how the various molecules interact. Statistical mechanics together with quantum mechanics have formed a basis for studying the physics of matter, i.e., the study from the atomic point of view of the various properties of matter. In general, the constituent molecules, even if we assume they obey the principles of dynamics, have extremely complicated means of motion.

At present, mathematically these motions are virtually impossible to categorize. So, one reasonably successful method around this problem has been to form an ideal realization of matter. This realization takes the form of a statistical mechanical model that is a simplified copy of matter. The pivot that is essential for the model to at least have mathematical meaning is a function Z called the *partition function*,

$$Z = \sum_{\sigma} \exp\left(\frac{-E(\sigma)}{kT}\right),$$

in which we define σ to be a *state* of the particular model, $E(\sigma)$ to be the total energy of this state, T to be the absolute temperature, and k to be Boltzmann's constant. The sum itself is taken over *all* the states of the particular model.

If the partition function of a model can be derived exactly, then this model is said to be *exactly solvable*. Numerous models have been shown to be exactly solvable, especially since the advent of Drinfel'd's quantum group. Due to this idea of a quantum group, and also by independent work in statistical mechanics, the partition function has shown to be closely related to invariants of knots (and links).

In classical two-dimensional statistical mechanics (i.e., statistical mechanics that applies classical dynamics to the dynamics of the microscopic world), two types of models have shown to be the most effective, the vertex model and the IRF (interaction round a face) model. These models not only allow us, in a very straightforward manner, to recover from the partition function the Jones polynomial, but in fact lead to a whole new series of Jones-type knot invariants [WAD].

In this chapter, we shall describe the exactly solvable 6-vertex model, which is the first in the above series of knot invariants. This model, or to be precise, the actual partition function, allows us to introduce a method that culminates in a different approach to the Jones polynomial. In the third, final section, we explain how to place statistical mechanical concepts and properties in a knot theoretical setting, which in turn allows a general method of constructing skein invariants to be introduced.

§1 The 6-vertex model

Let us consider a 2-dimensional lattice, Figure 12.1.1.

Figure 12.1.1

On the lattice we shall define a model called the *vertex model*. This model is one of the classical models of 2-dimensional statistical mechanics. As the name suggests, on the four edges that emanate from a vertex, of the vertex model, we assign *state variables*, i, j, k, l, Figure 12.1.2.

Figure 12.1.2

For this model it is possible to determine the realization probability, w, at a vertex. This probability is denoted by $w(i, j, k, l)(u)$ and is usually called a *Boltzmann weight*. The variable u is called the spectral parameter, and it indicates the mutual interactions of the system.

It is well known that for a statistical mechanical model to be solvable, the sufficient condition is the *Yang-Baxter equation*. In the case of the vertex model, this equation has the following form, and a diagrammatic interpretation is given in Figure 12.1.3:

$$\sum_{a,b,c} w(b, c, q, r)(u)w(a, k, p, c)(u + v)w(i, j, a, b)(v)$$

$$= \sum_{a,b,c} w(a, b, p, q)(v)w(i, c, a, r)(u + v)w(j, k, b, c)(u). \tag{12.1.1}$$

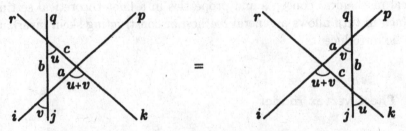

Figure 12.1.3

Therefore, if we can find a Boltzmann weight, w, that satisfies (12.1.1), the model is solvable, i.e., it is an exactly solvable model.

The 6-vertex model is an example of such an exactly-solvable model. In this case, the state variables can only take the value $\frac{1}{2}$ or $-\frac{1}{2}$, and at a vertex we assume conservation of the state variables, i.e., $i+j = k+l$, Figure 12.1.2. If the above conservation condition is not satisfied the Boltzmann weight at that vertex is zero.

Proposition 12.1.1.

The 6-vertex model has a set of Boltzmann weights $w(i, j, k, l)(u)$, *defined below, that satisfy the Yang-Baxter equation* (12.1.1) :

$$\left.\begin{array}{l} w(\tfrac{1}{2}, \tfrac{1}{2}, \tfrac{1}{2}, \tfrac{1}{2})(u) = w(-\tfrac{1}{2}, -\tfrac{1}{2}, -\tfrac{1}{2}, -\tfrac{1}{2})(u) = 1 \\[2mm] w(\tfrac{1}{2}, -\tfrac{1}{2}, \tfrac{1}{2}, -\tfrac{1}{2})(u) = w(-\tfrac{1}{2}, \tfrac{1}{2}, -\tfrac{1}{2}, \tfrac{1}{2})(u) = \dfrac{\sinh u}{\sinh(\lambda - u)} \\[2mm] w(\tfrac{1}{2}, -\tfrac{1}{2}, -\tfrac{1}{2}, \tfrac{1}{2})(u) = w(-\tfrac{1}{2}, \tfrac{1}{2}, \tfrac{1}{2}, -\tfrac{1}{2})(u) = \dfrac{\sinh \lambda}{\sinh(\lambda - u)} \end{array}\right\} \quad (12.1.2)$$

in all other cases $w(i, j, k, l)(u) = 0$.

In the above, $\sinh u$ is the hyperbolic sine function, namely, $\sinh u = \frac{1}{2}(e^u - e^{-u})$, and λ is another parameter.

In order for the Boltzmann weights to be more easily digested we shall rewrite them in the form of a 4×4 matrix, which we will denote by R. (Often in statistical mechanics, this matrix is called an *S-matrix*; however, it is more common, in the present context, due to another definition/extension of this same concept by Drinfel'd, to call it an *R-matrix*.) The columns of the R-matrix are indexed by the sets (i, j) and the rows by the sets (k, l), and the order is the reverse dictionary order. Excoriating the mathematical argot, the first column (and also the first row) is indexed by $(\frac{1}{2}, \frac{1}{2})$, the next column (row) by $(\frac{1}{2}, -\frac{1}{2})$, followed by $(-\frac{1}{2}, \frac{1}{2})$, and the final column (row) by $(-\frac{1}{2}, -\frac{1}{2})$. Therefore,

$$R = \|w(i, j, k, l)(u)\| = \begin{bmatrix} 1 & 0 & 0 & 0 \\ 0 & \frac{\sinh u}{\sinh(\lambda-u)} & \frac{\sinh \lambda}{\sinh(\lambda-u)} & 0 \\ 0 & \frac{\sinh \lambda}{\sinh(\lambda-u)} & \frac{\sinh u}{\sinh(\lambda-u)} & 0 \\ 0 & 0 & 0 & 1 \end{bmatrix}. \quad (12.1.3)$$

It is easy to see from this representation that if $i+j \neq k+l$, then $w(i, j, k, l)(u) = 0$.

In fact, these Boltzmann weights satisfy other important conditions in statistical mechanics besides the Yang-Baxter equation, for example,

$$w(i, j, k, l)(0) = \delta_{il}\delta_{jk}, \tag{12.1.4}$$

where δ_{pq} is the Kronecker delta symbol, which is 1 if $p = q$ and 0 otherwise. Also,

$$w(i, j, k, l)(u) = w(-i, -j, -k, -l)(u). \tag{12.1.5}$$

From the point of view of knot theory, these Boltzmann weights also satisfy the following important condition, called the *unitary condition*.

$$\sum_{p,q=\frac{1}{2},-\frac{1}{2}} w(i, j, p, q)(u)w(q, p, l, k)(-u) = \delta_{ik}\delta_{jl}. \tag{12.1.6}$$

If we depict this unitary condition (12.1.6), as shown in Figure 12.1.4, then it should immediately bring to mind the Reidemeister move, Ω_2 (if we ignore the under- and over-crossing information at the crossing points). Moreover, the Yang-Baxter equation (12.1.1) is essentially nothing but the Reidemeister move Ω_3. In other words, the partition function is unchanged by the regular moves Ω_2 and Ω_3. So, in essence, it should provide us with a *regular knot invariant*. Therefore, by Kauffman's principle (Chapter 11, Section 4) it should yield a knot invariant. So finding solutions to the Yang-Baxter equation seems to imply that we then may transform these solutions into knot invariants. In fact, this is not only feasible but in the late 1980s and early 1990s was undertaken with great gusto, mainly through the use of Drinfel'd's quantum group, [D]. However, to explain the quantum group approach would require a substantial amount of new notation and definitions without a significant increase of insight into knot theory. Therefore, we shall work within the more accessible framework of statistical mechanics. The interested reader might wish to consult [Tu].

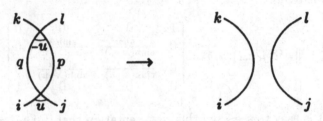

Figure 12.1.4

From the point of view of statistical mechanics, in order to fulfill the objective of deriving a knot invariant, we need to remove the parameters u and v that occur in (12.1.1) and (12.1.6), i.e., we need a set of Boltzmann weights that are "independent" of u and v. Hence, the necessary requirement is to find u and v in (12.1.1) such that $u = u + v = v$. The most obvious solution is to set $u = v = 0$. However, if we do this, then by (12.1.4) it follows that the only non-zero Boltzmann weights are $w(i, j, j, i)$, and these are equal to 1. So the R-matrix is

$$\begin{bmatrix} 1 & 0 & 0 & 0 \\ 0 & 0 & 1 & 0 \\ 0 & 1 & 0 & 0 \\ 0 & 0 & 0 & 1 \end{bmatrix},$$

and this will not lead to a new knot invariant. Another solution is to let $u, v \to \infty$. In this case what does the matrix in (12.1.3) of the subsequent Boltzmann weights look like? Let us do the necessary calculations (we assume λ is fixed):

$$w(\frac{1}{2}, -\frac{1}{2}, \frac{1}{2}, -\frac{1}{2})(\infty) = w(-\frac{1}{2}, \frac{1}{2}, -\frac{1}{2}, \frac{1}{2})(\infty)$$

$$= \lim_{u \to \infty} \frac{e^u - e^{-u}}{e^{\lambda - u} - e^{-(\lambda - u)}}$$

$$= \lim_{u \to \infty} \frac{1 - e^{-2u}}{e^{\lambda - 2u} - e^{-\lambda}}$$

$$= -e^{\lambda},$$

and

$$w(\frac{1}{2}, -\frac{1}{2}, -\frac{1}{2}, \frac{1}{2})(\infty) = w(-\frac{1}{2}, \frac{1}{2}, \frac{1}{2}, -\frac{1}{2})(\infty)$$

$$= \lim_{u \to \infty} \frac{\sinh \lambda}{\sinh(\lambda - u)}$$

$$= 0.$$

Therefore, the matrix in (12.1.3) now has the following form

$$\|w(i, j, k, l)(\infty)\| = \begin{bmatrix} 1 & 0 & 0 & 0 \\ 0 & -e^{\lambda} & 0 & 0 \\ 0 & 0 & -e^{\lambda} & 0 \\ 0 & 0 & 0 & 1 \end{bmatrix}. \tag{12.1.7}$$

However, this still does not lead to a very interesting invariant. So, in order to construct "non-trivial," from the point of view of knot theory, Boltzmann weights, we need to multiply $w(i,j,k,l)(u)$ by a *crossing multiplier* and let $u \to \infty$. The crossing multiplier we need is

$$e^{\frac{1}{2}(k-i-l+j)u},$$

and the subsequent Boltzmann weights are

$$\widetilde{w}(l,k,i,j)(u) = e^{\frac{1}{2}(k-i-l+j)u}w(i,j,k,l)(u).$$

(*Nota bene*, particular care needs to be taken with the order of i,j,k,l, see Figure 12.1.2.)

Finally, set

$$\widetilde{w}(i,j,k,l) = \lim_{u\to\infty} \widetilde{w}(i,j,k,l)(u).$$

For example,

$$\widetilde{w}(\frac{1}{2},-\frac{1}{2},\frac{1}{2},-\frac{1}{2}) = \lim_{u\to\infty} e^{-u}w(\frac{1}{2},-\frac{1}{2},-\frac{1}{2},\frac{1}{2})(u)$$

$$= \lim_{u\to\infty} e^{-u}\frac{\sinh\lambda}{\sinh(\lambda-u)}$$

$$= 0;$$

$$\widetilde{w}(-\frac{1}{2},\frac{1}{2},-\frac{1}{2},\frac{1}{2}) = \lim_{u\to\infty} e^{u}w(-\frac{1}{2},\frac{1}{2},\frac{1}{2},-\frac{1}{2})(u)$$

$$= \lim_{u\to\infty} \frac{e^{u}\sinh\lambda}{\sinh(\lambda-u)}$$

$$= 1 - e^{2\lambda}.$$

If we set $e^{2\lambda} = t$, then we can write the R-matrix of the (new) Boltzmann weights, $\widetilde{w}(i,j,k,l)$:

$$R = \|\widetilde{w}(i,j,k,l)\| = \begin{bmatrix} 1 & 0 & 0 & 0 \\ 0 & 0 & -\sqrt{t} & 0 \\ 0 & -\sqrt{t} & 1-t & 0 \\ 0 & 0 & 0 & 1 \end{bmatrix}. \qquad (12.1.8)$$

Exercise 12.1.1. Confirm the matrix in (12.1.8) by calculating the remaining $\widetilde{w}(i,j,k,l)$.

Since R is a square matrix, with a non-zero determinant, we can calculate its inverse matrix R^{-1} :

$$R^{-1} = \begin{bmatrix} 1 & 0 & 0 & 0 \\ 0 & 1-\frac{1}{t} & -\frac{1}{\sqrt{t}} & 0 \\ 0 & -\frac{1}{\sqrt{t}} & 0 & 0 \\ 0 & 0 & 0 & 1 \end{bmatrix}.$$

Let us denote the element $((i,j), (k,l))$ of R^{-1} by $\widetilde{w}_-(i,j,k,l)$; similarly, let us write as $\widetilde{w}_+(i,j,k,l)$ the element $((i,j), (k,l))$ of R. Although the $\widetilde{w}(i,j,k,l)(u)$ differ from the original $w(i,j,k,l)(u)$, the *sina qua non* for a knot invariant, the Yang-Baxter equation, (12.1.1), and (12.1.6) still hold. (The other conditions mentioned above may not hold.)

Exercise 12.1.2. Show that the Yang-Baxter equation holds for the Boltzmann weights $\widetilde{w}(i,j,k,l)(u)$.

Exercise 12.1.3. Show that the following formula holds:

$$\sum_{p,q=\frac{1}{2},-\frac{1}{2}} \widetilde{w}(i,j,p,q)(u)\widetilde{w}(p,q,k,l)(-u) = \delta_{ik}\delta_{jl}.$$

[Hint: Note that $\widetilde{w}(i,j,k,l)(u)$ is 0 if $i+j \neq k+l$.]

§2 The partition function for braids

In the previous section we took a lattice to be our model of "matter." A lattice without much scrutiny may be thought to be a braid. Hence, our objective in this section is to define the partition function of this model, i.e., a braid, using the R-matrix.

So, suppose β is a (oriented) n-braid and D is a regular diagram of β. At each crossing point of D, let us look at the four segments that make up a neighbourhood of that crossing point. We may assign a state s on the braid by placing a state variable $\frac{1}{2}$ or $-\frac{1}{2}$ on each of these four segments (see Figures 12.2.3 and 12.2.4) at each crossing point. For this given state, we may assign a Boltzmann weight at each crossing point of D, as described below.

On the four segments close to a crossing point, c, suppose the state variables are assigned as shown in Figure 12.2.1.

(a)   (b)

Figure 12.2.1

Then,

(i) if the crossing point, c, is positive, Figure 12.2.1(a), then assign $\widetilde{w}_+(l,k,i,j)$ to c;

(ii) if the crossing point, c, is negative, Figure 12.2.1(b), then assign $\widetilde{w}_-(l,k,i,j)$ to c.

Finally, for a fixed state, s, we take the product of *all* the Boltzmann weights, namely,

$$\prod_c \widetilde{w}_\pm(l,k,i,j). \tag{12.2.1}$$

We form a knot (or link) from a braid by adding closure strings, Figure 12.2.2.

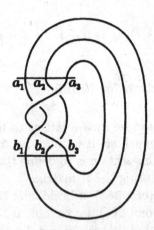

Figure 12.2.2

These closure strings will also have a contribution to a subsequent knot invariant. Hence, we need also to assign state variables to these closure strings. But if, for a given state, s, the state variable a_k is assigned to the top half of the kth closure string, and the state variable b_k to the bottom half of the kth closure string, then we shall assume they are equal, see Figure 12.2.2. By adding these closure strings, we no longer have a lattice model in the original sense, but a model with certain boundary conditions. Therefore, we need to perform "some sort of modification" to the product in (12.2.1). In fact, it is known that even for a statistical mechanical model with boundary conditions, a modification is required. The result is that for a knot (or link) K with a regular diagram D formed from a braid β and with an assigned state, s, we have the following modified function:

$$\prod_c \widetilde{w}_{\pm}(l, k, i, j) t^{-(a_1+a_2+\cdots+a_n)}, \qquad (12.2.2)$$

where $(a_1, \ldots, a_n)$ are the state variables that have been assigned to the top half of the closure strings of the braid β.

The factor $t^{-(a_1+a_2+\cdots+a_n)}$ that has been added is the "some sort of modification" that was alluded to previously. The product given in (12.2.2) is calculated separately for each state, s, on D. The sum (over all states) of these products is the *partition function* Z_β for this "matter" (i.e., the closed braid),

$$Z_\beta = \sum_s \prod_c \widetilde{w}_{\pm}(l, k, i, j) t^{-(a_1+a_2+\cdots+a_n)}. \qquad (12.2.3)$$

In order to calculate Z_β, usually it is not necessary to consider all the states, s, but rather only those for which the product in (12.2.2) is non-zero. Such states are called *contributing states*.

Let us now use the above partition function to calculate several examples with the Boltzmann weights of the R-matrix in (12.1.8).

Example 12.2.1. For the case $\beta = \sigma_1$, there are only three contributing states, as shown in Figure 12.2.3.

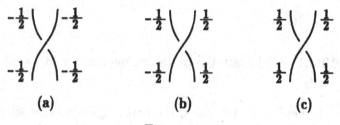

Figure 12.2.3

We have as the product of (12.2.2), respectively,

$$\text{(a)} \ \tilde{w}_+(-\frac{1}{2}, -\frac{1}{2}, -\frac{1}{2}, -\frac{1}{2})t^1 = t;$$

$$\text{(b)} \ \tilde{w}_+(-\frac{1}{2}, \frac{1}{2}, -\frac{1}{2}, \frac{1}{2})t^0 = 1 - t;$$

$$\text{(c)} \ \tilde{w}_+(\frac{1}{2}, \frac{1}{2}, \frac{1}{2}, \frac{1}{2})t^{-1} = t^{-1}.$$

Hence, by means of (12.2.3), the partition function is

$$Z_\beta = t + 1 - t + t^{-1} = 1 + t^{-1}.$$

Example 12.2.2. If $\beta = \sigma_1^2$, then the number of contributing states is 5, as shown in Figure 12.2.4.

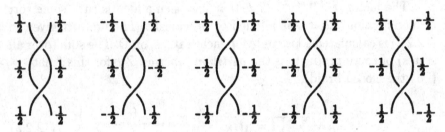

Figure 12.2.4

In a similar way as in the previous example, the partition function is

$$\begin{aligned}
Z_\beta &= \tilde{w}_+(\frac{1}{2}, \frac{1}{2}, \frac{1}{2}, \frac{1}{2})^2 t^{-1} + \tilde{w}_+(-\frac{1}{2}, \frac{1}{2}, \frac{1}{2}, -\frac{1}{2})\tilde{w}_+(\frac{1}{2}, -\frac{1}{2}, -\frac{1}{2}, \frac{1}{2})t^0 \\
&\quad + \tilde{w}_+(-\frac{1}{2}, -\frac{1}{2}, -\frac{1}{2}, -\frac{1}{2})^2 t + \tilde{w}_+(-\frac{1}{2}, \frac{1}{2}, -\frac{1}{2}, \frac{1}{2})^2 t^0 \\
&\quad + \tilde{w}_+(\frac{1}{2}, -\frac{1}{2}, -\frac{1}{2}, \frac{1}{2})\tilde{w}_+(-\frac{1}{2}, \frac{1}{2}, \frac{1}{2}, -\frac{1}{2})t^0 \\
&= t^{-1} + \left(-\sqrt{t}\right)^2 + t + (1-t)^2 + \left(-\sqrt{t}\right)^2 \\
&= (1 + t^{-1})(1 + t^2).
\end{aligned}$$

Exercise 12.2.1. Calculate the partition function for $\beta_1 = \sigma_1^3$ and $\beta_2 = \sigma_1\sigma_2^{-1}\sigma_1\sigma_2^{-1}$.

To find a "new" knot invariant, the first stage has been achieved, and we have a viable candidate in the partition function. However, to show that this leads to a knot invariant, we need

"the partition function to be equal for M-equivalent braids."

Without involving ourselves in unnecessarily messy definitions, the easiest approach is to associate a braid , via the Boltzmann weights (12.1.8), with some matrix. In the next section we shall show that then Z_β may be thought of as the trace of the matrix, and this trace is invariant under the Markov move M_1. Then by Kauffman's principle, if we multiply Z_β by a suitable factor, we shall have the "new" knot invariant. At this juncture we shall just introduce this knot invariant.

Theorem 12.2.1.

Suppose K is an (oriented) knot (or link) formed from a braid β, and that Z_β is the partition function for β. Then,

$$P_K(t) = t^{\frac{w(\beta)+1}{2}} Z_\beta \qquad (12.2.4)$$

is an invariant of K, where $w(\beta)$ is the Tait number of the regular diagram D of the closure of β.

For a closed braid the Tait number is very easy to calculate. Suppose $\beta = \sigma_{i_1}^{\varepsilon_1} \ldots \sigma_{i_m}^{\varepsilon_m}$ $(\varepsilon_j = \pm 1)$, then its Tait number is just the sum of its exponents, i.e., $w(\beta) = \varepsilon_1 + \varepsilon_2 + \ldots + \varepsilon_m$.

If we set

$$\widetilde{P}_K(t) = \frac{P_K(t)}{1+t},$$

then this is equivalent to the Jones polynomial of K. (When K is the trivial knot, then $P_K(t) = 1 + t$ [cf. Example 12.2.1 and (12.2.4)]; for this reason, we normalize $P_K(t)$ by the factor $1 + t$. In essence, there is no difference between $\widetilde{P}_K(t)$ and $P_K(t)$.)

Exercise 12.2.2. Calculate the partition function Z_β for the cases of $\beta_1 = \sigma_1 \sigma_2$ and $\beta_2 = \sigma_1 \sigma_2^{-1}$. Compare these to the partition function in Example 12.2.1. Also, determine $P_K(t)$ for these two braids.

Exercise 12.2.3. Prove that $\widetilde{P}_K(t)$ with regard to the skein diagrams D_+, D_- and D_0 satisfies

$$\frac{1}{t}\widetilde{P}_{D_+}(t) - t\widetilde{P}_{D_-}(t) = \left(\frac{1}{\sqrt{t}} - \sqrt{t}\right)\widetilde{P}_{D_0}(t), \qquad (12.2.5)$$

and further show that if K is a μ-component link, then

$$\widetilde{P}_K(t) = (-1)^{\mu-1}V_K(t).$$

§3 An invariant of knots

So far in this chapter we have concerned ourselves with the 6-vertex model, but there are infinitely many exactly solvable models. Using the Boltzmann weights from the various exactly solvable models, Wadati and his co-workers were able to discover an (infinite) series of invariants, which may be said to be a hierarchical extension of the Jones polynomial. In this section, we shall show in a systematic fashion how these skein invariants may be constructed.

Let us suppose, in what follows, that $N \geq 2$ is a positive integer and R is an $N^2 \times N^2$ invertible matrix. We may denote R as $R = \|R(i, j \mid k, l)\|$, where (i, j), (k, l) are chosen from the N^2 sets of pairs $(1, 1)$, $(1, 2)$, $\ldots$, $(1, N)$, $(2, 1)$, $\ldots$, (N, N) and (i, j) signifies the appropriate row of R and (k, l) the appropriate column of R. Suppose also the element $R(i, j \mid k, l)$ of R is an element of some ring Q. The ring Q can be arbitrary, with the sole proviso that it is a ring that contains t and $t^{\frac{1}{2}} (= \sqrt{t})$. The set of Laurent polynomials in $\sqrt{t}$ with rational coefficients is a typical example of Q. This matrix R may now be used to form $(r-1)$ $N^2 \times N^2$ matrices $R_i(r)$ $(i = 1, 2, \ldots, r-1)$.

This is done as follows. Suppose $r (\geq 2)$ is a positive integer, then with respect to $i = 1, 2, \ldots, r-1$, let

$$R_i = \underbrace{I \otimes \cdots \otimes I}_{(i-1) \text{ terms}} \otimes R \otimes \underbrace{I \otimes \cdots \otimes I}_{(r-i-1) \text{ terms}}, \qquad (12.3.1)$$

where I is the $N \times N$ identity matrix. The tensor product $A \otimes B$ of two matrices A and B is defined as follows. Suppose A is a $p \times p$ matrix of the form

$$A = \begin{bmatrix} a_{11} & a_{12} & \cdots & a_{1p} \\ a_{21} & a_{22} & \cdots & a_{2p} \\ \vdots & \vdots & \ddots & \vdots \\ a_{p1} & a_{p2} & \cdots & a_{pp} \end{bmatrix},$$

and similarly let B be a $q \times q$ matrix. Then $A \otimes B$ is a $pq \times pq$ matrix of the form

$$A \otimes B = \begin{bmatrix} a_{11}B & a_{12}B & \cdots & a_{1p}B \\ a_{21}B & a_{22}B & \cdots & a_{2p}B \\ \vdots & \vdots & \ddots & \vdots \\ a_{p1}B & a_{p2}B & \cdots & a_{pp}B \end{bmatrix}.$$

Example 12.3.1 Suppose

$$A = \begin{bmatrix} a_{11} & a_{12} \\ a_{21} & a_{22} \end{bmatrix} \quad \text{and} \quad B = \begin{bmatrix} b_{11} & b_{12} \\ b_{21} & b_{22} \end{bmatrix},$$

then

$$A \otimes B = \begin{bmatrix} a_{11}B & a_{12}B \\ a_{21}B & a_{22}B \end{bmatrix} = \begin{bmatrix} a_{11}b_{11} & a_{11}b_{12} & a_{12}b_{11} & a_{12}b_{12} \\ a_{11}b_{21} & a_{11}b_{22} & a_{12}b_{21} & a_{12}b_{22} \\ a_{21}b_{11} & a_{21}b_{12} & a_{22}b_{11} & a_{22}b_{12} \\ a_{21}b_{21} & a_{21}b_{22} & a_{22}b_{21} & a_{22}b_{22} \end{bmatrix}.$$

We shall index the rows (and columns) of R by the r-tuple positive integers $(a_1, a_2, \ldots, a_r)$, $1 \le a_1, a_2, \ldots, a_r \le N$, arranged in dictionary order. So the index ordering is as follows: $(1, 1, \ldots, 1)$, $(1, 1, \ldots, 2)$, $\ldots$, $(1, 1, \ldots, N)$, $(1, 1, \ldots, 2, 1)$, $\ldots$, $(1, 1, \ldots, 2, N)$, with the sequence continuing in this manner until the final indexing term $(N, N, \ldots, N)$ is reached.

Definition 12.3.1. For every $i = 1, 2, \ldots, r - 2$, if $R_i(r)$ satisfies the following condition

$$R_i(r)R_{i+1}(r)R_i(r) = R_{i+1}(r)R_i(r)R_{i+1}(r), \tag{12.3.2}$$

then $\{R_1(r), R_2(r), \ldots, R_{r-1}(r)\}$ are called *Yang-Baxter operators*.

These matrices $R_i(r)$, by definition, satisfy the following condition:

$$R_i(r)R_j(r) = R_j(r)R_i(r) \quad \text{if} \quad |i - j| \ge 2. \tag{12.3.3}$$

Therefore, due to (12.3.2) and (12.3.3), we have the correspondence

$$B_r \ni \sigma_i \longrightarrow R_i(r),$$

which associates an element of the braid group, B_r, to some matrix.

Let us now give some examples of such matrices R.

Example 12.3.2. Suppose $N = 2$, then

$$R = \|R(i, j \mid k, l)\| = \begin{bmatrix} 1 & 0 & 0 & 0 \\ 0 & 0 & -\sqrt{t} & 0 \\ 0 & -\sqrt{t} & 1-t & 0 \\ 0 & 0 & 0 & 1 \end{bmatrix}.$$

[Note: This matrix is the same as the one in (12.1.8).]

Example 12.3.3. It is possible to generalize the R-matrix of the previous example. Suppose $1 \le a, b, c, d \le N$, where $N \ge 2$, then

(1) if $a + b \ne c + d$, $R(a, b \mid c, d) = 0$;

(2) Suppose $m = a + b = c + d$,

(i) if $a - d = c - b < 0$, then $R(a, b \mid c, d) = 0$;

(ii) if $a - d = c - b \ge 0$ then

$$R(a, b \mid c, d) = (-1)^{a+c} t^{-\frac{1}{2}(ab+cd+N(k-m+2)-(k+m))}$$

$$\times \left[\frac{(t : a - 1)(t : N - d)}{(t : k)(t : d - 1)(t : N - a)} \frac{(t : c - 1)(t : N - b)}{(t : k)(t : b - 1)(t : N - c)} \right]^{\frac{1}{2}},$$

where $k = a - d = c - b$, and the term $(t : n) = (1-t)(1-t^2) \dots (1-t^n)$, if n is a positive integer, and equal to 1 if $n = 0$.

The above Boltzmann weights are basically the same as those obtained by Wadati and his co-workers from the Boltzmann weights of the N-vertex model.

Example 12.3.4. Suppose $1 \le i, j, k, l \le N$, then

(1) if $i = j = k = l$, $\quad R(i, j \mid k, l) = -t$;

(2) if $i = l \ne k = j$, $\quad R(i, j \mid k, l) = 1$;

(3) if $i = k < j = l$, $\quad R(i, j \mid k, l) = t^{-1} - t$;

(4) in all other cases, $\quad R(i, j \mid k, l) = 0$.

Therefore, $R(i, j \mid k, l) \ne 0$ only if the condition $\{i, j\} = \{k, l\}$ holds.

Example 12.3.5. Suppose $N = 2$, then we may set

$$R = \|R(i, j \mid k, l)\| = \begin{bmatrix} 1 & 0 & 0 & 0 \\ 0 & 0 & -\sqrt{t} & 0 \\ 0 & -\sqrt{t} & 1 - t & 0 \\ 0 & 0 & 0 & -t \end{bmatrix}.$$

All four of the above examples are Yang-Baxter operators, and hence from them we may define invariants of knots (and links). In particular, in the final example (Example 12.3.5) the question of how we define a non-trivial invariant is of immediate interest since this invariant is zero for *all* knots and links (see Exercise 12.3.6).

So, how exactly do we define an invariant of knots (or links) from a given set of Boltzmann weights? First of all, we must describe the partition function in terms of the Yang-Baxter operators, and then, as in the previous section, we shall need "some sort of modification" of the partition function. The necessary "modification" corresponds to a $N \times N$-diagonal matrix μ :

$$\mu = \begin{bmatrix} \mu_1 & & & & \\ & \mu_2 & & O & \\ & & \ddots & & \\ & O & & \mu_{N-1} & \\ & & & & \mu_N \end{bmatrix},$$

where μ_i is a non-zero element of Q.

Definition 12.3.2. Suppose a and b are non-zero invertible elements of Q. Then if the set $\{R, \mu, a, b\}$ satisfies the conditions in (12.3.4), it is called an *enhanced Yang-Baxter operator (or matrix)*.

(1) For $1 \le i, j, k, l \le N$, $(\mu_i \mu_j - \mu_k \mu_l) R(i, j \mid k, l) = 0.$

(2)   (i) $\displaystyle\sum_{j=1}^{N} R(i, j \mid k, j) \mu_j = ab\delta_{ik};$

$$\text{(12.3.4)}$$

(ii) $\displaystyle\sum_{j=1}^{N} R^{-1}(i, j \mid k, j) \mu_j = a^{-1} b\delta_{ik}.$

The (homomorphic) map φ_R, which, if we recall, sends a generator σ_i of the r-braid group B_r to $R_i(r)$, allows us to represent an arbitrary element, $\beta = \sigma_{j_1}^{\varepsilon_1} \sigma_{j_2}^{\varepsilon_2} \dots \sigma_{j_m}^{\varepsilon_m}$ of B_r, by an $N^r \times N^r$ matrix, i.e.,

$$\varphi_r(\beta) = R_{j_1}^{\varepsilon_1}(r) R_{j_2}^{\varepsilon_2}(r) \dots R_{j_m}^{\varepsilon_m}(r).$$

This product is "modified" (multiplied) by the $N^r \times N^r$ matrix given by

$$\mu^{(r)} = \underbrace{\mu \otimes \mu \otimes \dots \otimes \mu}_{r \text{ times}}.$$

The final operation required to define a knot (or link) invariant is to take the trace, i.e., $\operatorname{tr}(\varphi_r(\beta) \mu^{(r)})$.

Theorem 12.3.1.
Suppose K is a (oriented) knot (or link) that is represented by the r-braid β, i.e., K is the closure of β. Then if a $(\ne 0)$ and b $(\ne 0)$ are

the elements of Q defined above (Definition 12.3.2), then the following
is an invariant of K:

$$J_\beta = a^{-w(\beta)} b^{-r} \mathrm{tr}(\varphi_r(\beta)\mu^{(r)}),$$

where $w(\beta)$ is the Tait number of β.

Let us denote J_β by J_K. Also, if $J_O = b^{-1}\mathrm{tr}(\mu)$ is not zero,
i.e., J_K of the trivial knot O is not zero, we can normalize J_K in the
following way:

$$\widehat{J}_K = \frac{J_K}{J_O}.$$

Exercise 12.3.1. Show that if $\alpha \underset{M}{\sim} \beta$ (Definition 10.3.2), then $J_\alpha = J_\beta$, and hence prove J_K is an invariant of K.

Therefore, to find a knot invariant by the above method, the important fact is to find an $N^2 \times N^2$ matrix R that is a Yang-Baxter operator. We have already found such Yang-Baxter operators in Examples $12.3.2 \sim 12.3.5$. The question now is, To what type, if any, of the previous knot (skein) invariants are they related to?

Example 12.3.2 (continued). If we set

$$a = t^{-\frac{1}{2}}, \ b = t^{\frac{1}{2}} \quad \text{and} \quad \mu = \begin{bmatrix} 1 & 0 \\ 0 & t \end{bmatrix},$$

then $\{R, \mu, a, b\}$ is an enhanced Yang-Baxter operator. It easily follows that if $i + j \neq k + l$, then $R(i, j \mid k, l) = 0$; and if $i + j = k + l$, then $\mu_i\mu_j - \mu_k\mu_l = 0$. So condition (1) of (12.3.4) is satisfied. We can calculate directly the appropriate equations of condition (2) in (12.3.4),

$$R(1,1 \mid 1,1)\mu_1 + R(1,2 \mid 1,2)\mu_2 = 1 + 0 = 1 = ab$$
$$R(2,1 \mid 2,1)\mu_1 + R(2,2 \mid 2,2)\mu_2 = (1 - t) + t = 1 = ab.$$

The calculation for (ii) of this condition is completely analogous to the above. In fact, the invariant $\widehat{J}_K$ that is derived from this enhanced Yang-Baxter operator is nothing other than the Jones polynomial.

Exercise 12.3.2. Show $\widehat{J}_K = \widetilde{P}_K(t)$, where $\widehat{J}_K$ is as in Example 12.3.2 and $\widetilde{P}_K(t)$ is as in Exercise 12.2.3.

Exercise 12.3.3. Use the Yang-Baxter operator in Example 12.3.2 to calculate J_β, by means of the trace, for $\beta = \sigma_1^2$.

Example 12.3.3 (continued). If we set $\mu_i = t^{i-1}$ $(i = 1, 2, \ldots, N)$ and $a = t^{-\frac{N-1}{2}}$, $b = t^{\frac{N-1}{2}}$, then $\{R, \mu, a, b\}$ is an enhanced Yang-Baxter operator, and for each $N = 2, 3, \ldots$, we obtain a knot invariant of the form,

$$\widehat{J}_K^{(N)} = \frac{(t^{\frac{N-1}{2}})^{w(\beta)-r+1} \operatorname{tr}(\varphi_r(\beta)\mu^{(r)})}{1 + t + \cdots + t^{N-1}},$$

where K is a knot (or link) that has been represented by the r-braid β. Moreover, $J_O^{(N)} = t^{-\frac{N-1}{2}}(1 + t + \cdots + t^{N-1})$. We leave it as a straightforward exercise for the reader to show that if $N = 2$, then this knot invariant is the same as the Jones polynomial, so it is appropriate to call $\widehat{J}_K^{(N)}$ the Nth degree Jones polynomial.

As an example of one of the polynomials that can be calculated, let $N = 3$ and K be the right-hand trefoil knot, then

$$\widehat{J}_K^{(3)} = t^2 + t^5 - t^7 + t^8 - t^9 - t^{10} + t^{11}.$$

Exercise 12.3.4. For the Boltzmann weights in Example 12.3.3, with $N = 3$, determine the 9×9 matrix R. Using this R-matrix calculate the 3rd degree Jones polynomial of the (oriented) Hopf link, Figure 4.3.2(c).

Example 12.3.4 (continued). If we set $\mu_i = t^{2i-N-1}$ and $a = -t^N$, $b = 1$, then $\{R, \mu, a, b\}$ is an enhanced Yang-Baxter operator. In this case, J_K of the trivial knot is

$$J_O = \frac{t^N - t^{-N}}{t - t^{-1}},$$

and

$$J_\beta = (-t^N)^{-w(\beta)} \operatorname{tr}(\varphi_r(\beta)\mu^{(r)})$$

is an invariant of a knot (or link) K that has been represented by the r-braid β.

By considering the R-matrices, it may be shown that $\widehat{J}_K = \frac{J_K}{J_O}$ satisfies the following skein relation:

$$t^N \widehat{J}_{D_+} - t^{-N} \widehat{J}_{D_-} = (t - t^{-1})\widehat{J}_{D_0}.$$

Since t and N are independent of each other, we may think of t^N as a distinct variable, then from the infinite series $\widehat{J}_2, \widehat{J}_3, \ldots$, we can *recover* the skein polynomial $P_L(v, z)$.

Exercise 12.3.5. Using the enhanced Yang-Baxter operator of Example 12.3.4, calculate J_β for $N = 3$ and $\beta = \sigma_1^2$.

Example 12.3.5 (continued). If we set

$$a = t^{\frac{1}{2}}, \ b = t^{-\frac{1}{2}} \ \text{ and } \ \mu = \begin{bmatrix} 1 & 0 \\ 0 & -1 \end{bmatrix},$$

then $\{R, \mu, a, b\}$ is an enhanced Yang-Baxter operator. However, for this enhanced Yang-Baxter operator, irrespective of the braid β chosen, $J_\beta = 0$. Also, since $J_O = b^{-1}\text{tr}(\mu) = 0$, $\widehat{J}_\beta = \frac{0}{0}$, i.e., the "knot invariant" cannot be determined from the methods described above. However, by delving a bit deeper (but not much deeper) into the theory this chapter is based on, it is possible to define a non-zero knot invariant from this enhanced Yang-Baxter operator and then show that it is equivalent to the Alexander polynomial.

Exercise 12.3.6. Show, using the enhanced Yang-Baxter operator of Example 12.3.5, that for $\beta = \sigma_1^2$, $J_\beta = 0$.

These knot invariants, which depend on the R-matrices that satisfy the Yang-Baxter equation, gave rise to a deluge of research into this area. In this chapter we have only been able to provide an introduction into exactly-solvable models and only the basics of the subsequent Jones-type invariants. The interested reader may wish to refer, amongst others, to Jones [J2] and Turaev [Tu].

It should be noted that there exist methods that allow us to calculate the above invariants directly from an arbitrary regular diagram of a knot (or link), rather than as we have done in this chapter from a braid. Finally, it should be underlined that these invariants are very closely related to the new invariants for closed orientable 3-manifolds [L], [Wi].

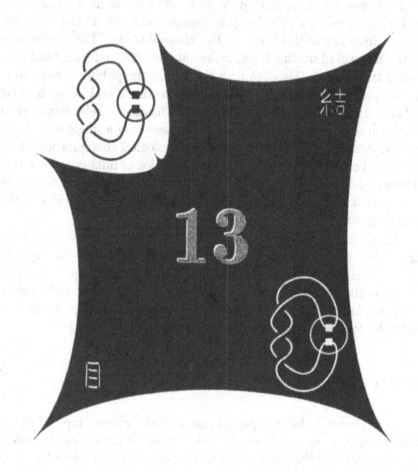

Knot Theory in Molecular Biology

F.H.C. Crick and J.D. Watson, in one of the most remarkable insights of the 20th century, unraveled the basic structure of DNA. For this profundity into the substance of living matter, they were jointly awarded the Nobel Prize for Medicine in 1962. Essentially, a molecule of DNA may be thought of as two linear strands intertwined in the form of a double helix with a linear axis. A molecule of DNA may also take the form of a ring, and so it can become tangled or knotted. Further, a piece of DNA can break temporarily. While in this broken state the structure of the DNA may undergo a physical change, and finally the

DNA will recombine. In fact, in the early 1970s it was discovered that a single enzyme called a (DNA) topoisomerase can facilitate this complete process, from the initial break to the recombination. The reader who might have picked up this book, looked at the title, and then randomly opened the book at this page may think that the publisher has somehow inserted some pages of an elementary textbook on biology here by mistake. But, let us reconsider the above. The double-helix structure of DNA – on some occasions DNA may even have only a single strand – is a geometrical entity, or more precisely, a topological configuration. This topological configuration is itself a manifestation of linking or knotting. Further, it has been shown when a topoisomerase causes DNA to change its form that the process is very similar to what happens locally in the skein diagrams.

Therefore, for the geometrical entity – knotted or linked – the linking number is an important concept, while the action of the topoisomerase is related to the new skein invariants. In this chapter we shall give an outline of exactly how knot theory is interpreted and used in trying to understand the changes DNA undergoes; this is sometimes called the *topological approach to enzymology*.

§1 DNA and knots

Living matter, be it a person, an animal, or some type of plant, *et cetera*, is composed of countless molecules. Within these countless molecules, we find the DNA of the living matter. The nature of a living thing and how it develops depends largely on the information it inherits from its DNA. Technically, it is the genes that carry the information of the DNA and are passed on from the progenitor to its offspring.

In general, as we have already mentioned, DNA has the structure of two linear strands intertwined along a linear axis, forming a double helix. This, however, is the not only possible structure for DNA, and in what follows the next description is probably more easy to comprehend in the context of knot theory. On some occasions, it has been found that DNA has the form of a *ring* consisting of either a single strand or two strands coiled in a double helix. This single-strand DNA can literally be knotted, i.e., the objects (knots and links) we have so far discussed, to a degree in a staid abstract way, can *actually* be seen under an electron microscope. The information the DNA molecule carries, i.e., the arrangement of its nucleotide base pairs, is unrelated to how it is knotted (or tangled). So, maybe we should dismiss the knot (or link) as a useful tool in molecular biology, but without much significance. However, recent research has shown that the knot type (of a DNA molecule) has an important effect on the actual function of the DNA molecule in the cell. Therefore, using knot theory techniques, it may be possible to bring further insight into the structure of a DNA molecule.

Let us now be bit more precise and describe a DNA molecule purely mathematically. A mathematical model for a DNA molecule is usually a thin, long, narrow (oriented) twisted ribbon, Figure 13.1.1. (In this figure, the ribbon is homeomorphic to $S^1 \times [-1, 1]$, but not to the Möbius band.)

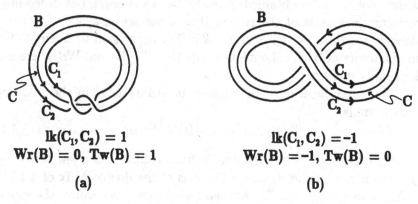

$$\text{lk}(C_1, C_2) = 1$$
$$\text{Wr}(B) = 0, \text{Tw}(B) = 1$$

(a)

$$\text{lk}(C_1, C_2) = -1$$
$$\text{Wr}(B) = -1, \text{Tw}(B) = 0$$

(b)

Figure 13.1.1

The two curves C_1 and C_2 that form the boundaries of the ribbon B represent the closed DNA strands. We may fix an orientation on the curve C that forms an axis for B (i.e., the central curve $S^1 \times \{0\}$). This orientation on C induces similar orientations on C_1 and C_2, on the boundary of B. In fact, the linking number between C_1 and C_2 $[\text{lk}(C_1, C_2)]$ is an invariant, and its change has a very important effect on the structure of the DNA molecule. For example, it is known that if we reduce the linking number of a double-strand DNA molecule, then the effect is to cause the DNA molecule to twist and coil, i.e., what is known as *supercoiling*. The actual reduction of the linking number of DNA molecule can be caused by a topoisomerase acting on the DNA molecule. (In fact, the orientation on an actual DNA molecule is not precisely as above. The orientations, in reality, on the two DNA strands are mutually opposite. Therefore, maybe we should assign mutually opposite orientations to C_1 and C_2. However, in the case of a link formed from C_1 and C_2, the linking number defined on the DNA molecule in biology, and the linking number calculated from the mathematical model, as above, are equal. So, from a numerical point of view, there is no incongruity.)

The number of twists the ribbon B has along the axis C is called the *twisting number,* and is denoted by $\text{Tw}(B)$. The writhe, $\text{Wr}(B)$, in the case of mathematical biology differs slightly from our previous definition, Definition 4.5.2. For the purposes of this chapter, we shall define the writhe as the *average* value of the sum of the signs of the crossing points, averaged over all the projections. Succinctly, the writhe is determined from the axis C by considering it as a spatial curve. These numbers, $\text{Wr}(B)$ and $\text{Tw}(B)$, are invariants. They are not, however, invariants of the knot (or link) obtained from the DNA molecule, but differential geometry invariants of the ribbon B as a surface in space. [If C is a plane curve, then $\text{Wr}(B) = 0$. So, $\text{Wr}(B)$ may be said to calculate the non-planarity of B. We should also note that $\text{Tw}(B)$ and $\text{Wr}(B)$ are not necessarily integers.]

The three "invariants" mentioned above are related by the following basic formula.

$$\text{lk}(C_1, C_2) = \text{Tw}(B) + \text{Wr}(B). \tag{13.1.1}$$

On occasions – when the double helix is unwound a few turns due to a cut in one of the strands – the axis of the double helix of a DNA molecule twists into a helix. As mentioned above, this causes the super-coiling of the DNA, and (13.1.1) is very useful in picking up this quality. Although supercoiling is interesting in its own right, we shall, to avoid having to swathe the reader in concepts from molecular biology, not delve any deeper into this concept. If the reader would like to pursue or become acquainted with these concepts, a good reference is Wang [Wa].

Thankfully, DNA is very malleable, being able to recombine through a series of phases; otherwise the world would be populated by clones. In the phases of this process, the knot type of the DNA molecule is actually changed. At first, it might seem that to understand this process from the point of view of molecular biology will be complicated. However, in the early 1970s it was found the whole process, from the original splicing to the recombination, was the result of the effect of a single enzyme/catalyst called a *topoisomerase.* The term *topoisomerase* may seem rather strange, but it is relatively easy to explain. Chemically, two molecules with the same chemical composition but different structure are called *isomers.* It follows that two DNA molecules with the same sequence of base pairs but different linking numbers are also *isomers.* Due to the difference in linking numbers, "topologically" they are inequivalent. So, these DNA molecules are called *topoisomers*. Hence, the enzyme that causes the linking number to change is termed an *topoisomerase*. The process of mutation due to a topoisomerase can be in simple terms be described as follows: First a strand of the DNA is cut at one place, then a segment of DNA passes through this cut, and finally the DNA reconnects itself.

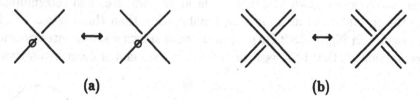

(a)   (b)

Figure 13.1.2

In Figure 13.1.2, we give two examples of the action of a topoisomerase on a DNA molecule (for clarity, we have not drawn the helical twist). The place where the strand is cut is denoted by "*o.*" The two figures [Figures 13.1.2(a) and (b)] are relatively self-explanatory. The single strand, in Figure 13.1.2(a), has a single cut due to a topoisomerase and the DNA passes through it and recombines; this is called a *Type I* topoisomerase. While, in Figure 13.1.2(b), a cut in a double-strand DNA, due again to a topoisomerase, allows a double-strand DNA to pass through it and recombine, this is as expected called a *Type II* topoisomerase. Finding such a topoisomerase is relatively straightforward, since they occur in organisms small and large, from bacteria to within the reader of this book.

In the next few sections we shall discuss in slightly more detail the effect of a certain topoisomerase (to be precise, it should really be called a *recombinase*). This effect is usually called a *site-specific recombination*.

§2 Site-specific recombination

As the name suggests, a site-specific recombination is a local operation. The effect of the recombinase on a DNA molecule is to either move a piece of this DNA molecule to another position within itself or to import a foreign piece of a DNA molecule into it. The result is that the gene transmutes itself. It is known, in fairly advanced organisms, of which we are an example, that various antibodies form through such site-specific recombination of a DNA molecule.

The exact process of a site-specific recombination is fairly easy to understand. Firstly, two points of the same or different DNA molecules are drawn together, either by a recombinase or by random (thermal) motion (or even possibly both). The recombinase then sets to work, causing the DNA molecule to be cut open at two points on the parts that have been drawn together. The loose ends are then recombined by the recombinase in a different combination than the original DNA molecule. In Figure 13.2.1(a) ∼ (c), we have shown a simple site-specific recombination that has been carried out in the manner described above.

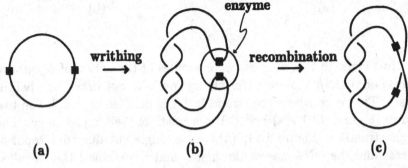

Figure 13.2.1

The above description is loosely what occurs in a site-specific recombination. For the reader who might want to read further and more precisely, we shall define the relevant terms involved in this process found in literature on this subject. The DNA molecule before the action of the recombinase is called a *substrate*; after the recombination it is called a *product*. The process of going from the DNA molecule to a state in which two parts of the DNA molecule have been drawn together, Figures 13.2.1(a) up to just before (b), is said to be the *writhing process*. When at this stage the recombinase combines with the substrate, the resultant combined complex is called a *synaptic complex*, Figure 13.2.1(b). Within the synaptic complex, we can assign local orientations to the respective, relatively small parts of the DNA molecule (or molecules) on which the recombinase acts [within the circle in Figure 13.2.2(a), (b), and (c)].

If the orientations on the DNA molecule *and* the orientation induced by these local orientations agree, then this arrangement is called a *direct repeat*, Figure 13.2.2(a). On the other hand, if they do not agree, then the arrangement is said to be an *inverted repeat*, Figure 13.2.2(c).

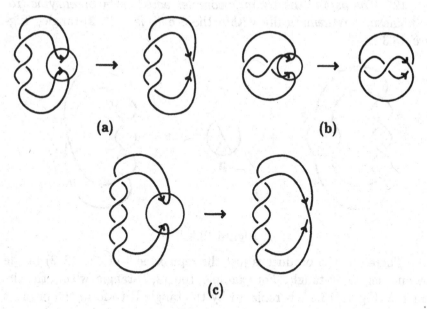

(a)   **(b)**

(c)

Figure 13.2.2

Exercise 13.2.1. For a site-specific recombination show the following:

(1) If the substrate is a DNA *knot* and the arrangement is a direct repeat, then the product is a 2-component DNA link, Figure 13.2.2(a). If, however, the arrangement is an inverted repeat, then the product is a DNA knot, Figure 13.2.2(c);

(2) If the substrate is a DNA *link* (i.e., two DNA molecules entwined), then after recombination the product is a DNA knot, Figure 13.2.2(b).

In the next section, we shall describe a mathematical model, with empirical constraints, for the site-specific recombination due to the action of a recombinase.

§3 A model for site-specific recombination

The following proposition follows from empirical evidence:

Proposition 13.3.1.

(1) *Almost all the products obtained by the site-specific recombination of trivial knot substrates are rational knots (or links), i.e., 2-bridge knots (or links).*

(2) *The part of the synaptic complex acted on by an enzyme (re-combinase), mathematically within the 3-ball, is a* (2, 2)-*tangle, Figure 13.3.1.*

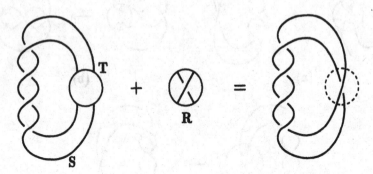

Figure 13.3.1

Therefore, the product is just the replacement of one (2, 2)-tangle by another (2, 2)-tangle. For example, the (2, 2)-tangle within the circle T in Figure 13.3.1 is replaced by the tangle R to form the product shown. This process may be expressed by means of our definition of the sum of tangles (cf. Chapter 9, Section 1). The good thing about mathematics is that inside may be outside, and outside may be inside. Mathematically, it is perfectly reasonable to consider S to be a (2, 2)-tangle in T. The numerator of the sum of S and R is then the product, Figure 13.3.2.

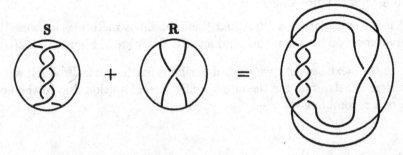

Figure 13.3.2

So the following "equation" holds.

$$N(S + R) = \text{the product.} \qquad (13.3.1)$$

Further, we may divide the substrate into the external tangle S and the internal tangle E, since the substrate is then the numerator of the sum of S and E, Figure 13.3.3.

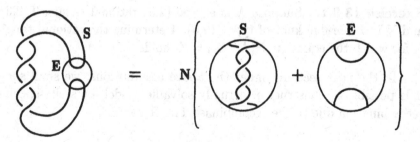

Figure 13.3.3

Again we have a quasi-equation holding,

$$N(S + E) = \text{the substrate.} \tag{13.3.2}$$

If it is possible to observe the substrate and the product, then the ideal situation would be to determine S, E, R from the two quasi-equations (13.3.1) and (13.3.2). Mathematically, however, since there are *only* two equations but three "unknowns," it is not possible without further assumptions to determine these unknowns:

So, we need to fall back on experimental data to make some further progress. Recently, the following has been observed:

Supposition 13.3.2.

The effect of the enzyme – the change due to this enzyme from the tangle E to the tangle R – depends only on the original enzyme, so this process of change is independent of the shape, position, and size of the substrate.

For each recombination we shall obtain a quasi-equation as above. However, by Supposition 13.3.2 no new indeterminates are added, and the indeterminate is always R. Therefore, even although the number of equations increases, the number of indeterminates remains constant. This means, mathematically, that there is a possibility that we can solve the *collective equations*. Another assumption from experimental observation is that the repetition of site-specific recombinations can be expressed as the sum of tangles.

Supposition 13.3.3.

The product of a series of site-specific recombinations can be expressed as the numerator of the sum of tangles, namely, it is of the form

$$N(S + R + R + \cdots + R).$$

Exercise 13.3.1. Suppose A is a type (2,3) rational tangle, T(2,3), and K is a 2-bridge knot of type (19,5). Determine the rational tangle X for which the equation $N(A + X) = K$ holds.

In the next section, under the above assumptions, we shall show it is possible to construct a virtually solvable model for a site-specific recombination due to the recombinase Tn3 Resolvase.

§4 Recombination due to the recombinase Tn3 Resolvase

As already mentioned, Tn3 Resolvase is an enzyme (recombinase) that is a catalyst for a site-specific recombination on a circular DNA substrate with directly repeated recombination sites. When this resolvase acts on a circular DNA substrate that is supercoiled and unknotted, then the product is a link. In most cases, the product is the Hopf link, Figure 13.4.1(a). (In addition, if the orientation of the DNA molecule is taken into account, then the linking number of the recombined DNA molecule may be considered to be −1. In the sequel, this fact will not be of relevance, but see Theorem 13.4.2.)

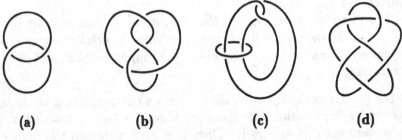

| (a) | (b) | (c) | (d) |

Figure 13.4.1

If the resolvase causes a further recombination, then the subsequent product is the figure 8 knot, Figure 13.4.1(b). Continuing, a further recombination (so three recombinations have occurred) produces the Whitehead link as the product, Figure 13.4.1(c). Up to three recombinations due to Tn3 resolvase have been shown experimentally to agree with the above. By experimental observation it has also been shown that the product of the fourth recombination is the knot in Figure 13.4.1(d). However, to find the original S, E, and R, we shall show that this fourth recombination is not required and will only be used as a check for the model we shall put forward.

Assuming Suppositions 13.3.2 and 13.3.3 hold, we can draw the series of diagrams in Figure 13.4.2.

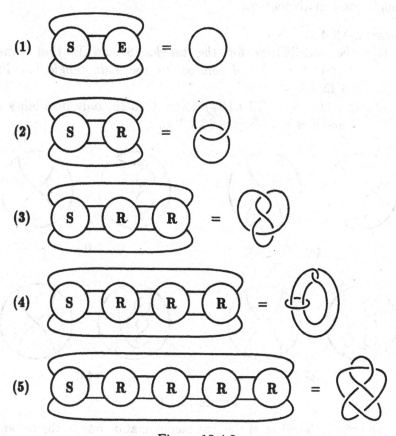

Figure 13.4.2

Since the knots and links on the right-hand side of the "equations" in Figure 13.4.2 are 2-bridge knots or links, we may rewrite them as mathematical formulae using the notation created in Chapter 9, Section 3:

(1) $N(S + E) = C(1)$
(2) $N(S + R) = C(2)$
(3) $N(S + R + R) = C(2, 1, 1)$ (13.4.1)
(4) $N(S + R + R + R) = C(1, 1, 1, 1, 1)$
(5) $N(S + R + R + R + R) = C(1, 1, 1, 2, 1)$.

We can, by looking carefully at these "equations," determine the tangles R and S. (It is not, however, possible to determine E from these equations; see Theorem 13.4.2). So, finally in this chapter we are ready to apply some mathematics.

Theorem 13.4.1.

    (1) *The possibilities for the tangles S and R that satisfy (13.4.1)(1) ~ (3) are limited to the four tangles in Figure 13.4.3.*

    (2) *In addition, if (13.4.1)(4) holds, then the only possibility for R and S is as in Figure 13.4.3(a).*

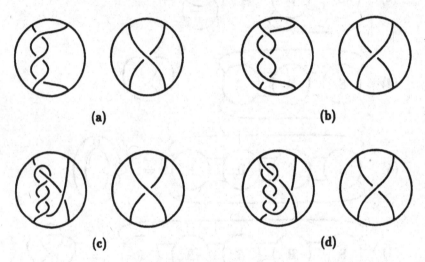

(a)   (b)

(c)   (d)

Figure 13.4.3

Therefore, the effect of the first recombination due to the recombinase Tn3 may be thought of as that in Figure 13.4.4.

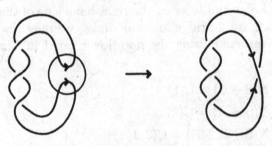

Figure 13.4.4

Since we cannot as yet determine E, the above recombination is to a certain degree not precise; however, if we assume E is the (0) tangle, the above results are known to hold.

<u>Proof</u>

The first step is for the reader to make sure, by drawing the relevant diagrams, the above tangles (Figure 13.4.3) are solutions to (13.4.1)(1) ~ (3). We shall give an outline of a proof showing these are the only possible solutions. (For a detailed proof refer to Ernst and Sumners [ES2].)

We can prove both R and S are rational tangles by calling on Theorem 9.3.1 and Proposition 9.3.4, in which we proved that a 2-bridge knot (or link) can be represented as a denominator or numerator of a rational tangle. (The proof itself is not very straightforward; so in order not to make this proof too dense, we shall omit the details of this part. In molecular biology, as a first assumption R and S are taken to be rational tangles.)

Next, let us determine from equations (13.4.1)(2) and (3) the rational tangles S and R. By Theorem 9.2.2, we know S and R may be represented, respectively, by the fractions $\frac{a}{b}$ and $\frac{c}{d}$ (recall, $\infty = \frac{1}{0}$).

Exercise 13.4.1. It is known that if both $\frac{a}{b}$ and $\frac{c}{d}$ are not integers, then $N(S+R+R)$ is *not* a rational knot (or link). Confirm this is the case for $\frac{a}{b} = \frac{7}{3}$ and $\frac{c}{d} = -\frac{5}{2}$.

So, if both R and S correspond, respectively, to integers [i.e., $b = d = 1$), $N(S+R+R) = N(T(a+2c))$], which is a torus knot (or link) of type $(a+2c, 2)$. This implies that the resultant knot cannot be the figure 8 knot. (This may be shown by comparing the Alexander polynomials.) Therefore, only one of S and R may correspond to an integer.

So, now, suppose S is a $(0,0)$-tangle (i.e., $b = 0$), then (see also Figure 13.4.5)

$$N(S+R+R) = N(T(0,0)+R+R) = D(R+R).$$

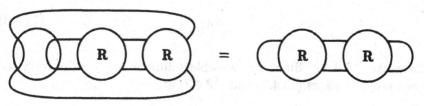

Figure 13.4.5

This also cannot be the figure 8 knot (why?). Moreover, if R is a $(0,0)$-tangle, then $N(S + R + R)$ is at the very least a 2-component link, and thus obviously not a knot, see Figure 13.4.6.

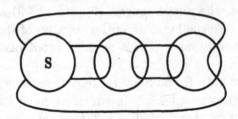

Figure 13.4.6

Therefore, neither R nor S may be a $(0,0)$-tangle.

So, let us assume R is an integer tangle, $T(r)$. If $r = 0$, then $N(S + R) = N(S)$, and similarly $N(S + R + R) = N(S)$. Therefore, $N(S + R) = N(S + R + R)$, which directly contradicts $(13.4.1)(2)$ and (3). The consequence of this is that r cannot be equal to zero. Hence, if we suppose $R = T(r)$ $(r \neq 0)$, then S must correspond to a rational tangle $\frac{u}{v}$. It is now possible to determine r, u, v by making use of Theorem 9.3.5.

Firstly, from $N(S+R) = C(2)$, the absolute value of the numerator of

$$r + \frac{u}{v} = \frac{rv + u}{v}$$

is equal to the determinant of $C(2)$, i.e.,

$$|rv + u| = 2. \tag{13.4.2}$$

Similarly, from $N(S + R + R) = C(2, 1, 1)$, the absolute value of the numerator of

$$\frac{u}{v} + 2r = \frac{u + 2rv}{v}$$

is equal to the determinant of $C(2,1,1)$. So, in this case it follows that

$$|u + 2rv| = 5. \tag{13.4.3}$$

Exercise 13.4.2. Show the possible solutions for r, u, v from the system of equations $(13.4.2)$ and $(13.4.3)$ are

$$\{(u, rv)\} = \{(-1, 3), (1, -3), (9, -7), (-9, 7)\}.$$

Let us look at the first set in these solutions, $u = -1$ and $rv = 3$. Since r and v are integers, we have the following possible solutions:

$$\left\{\left(\frac{u}{v}, r\right)\right\} = \left\{\left(-\frac{1}{3}, 1\right), \left(\frac{1}{3}, -1\right), (-1, 3), (1, -3)\right\}.$$

Since $\frac{u}{v}$ is not an integer, we may remove from our considerations the final two solutions. Therefore, in the case $\frac{u}{v} = -\frac{1}{3}$, the corresponding tangle is $T(-3, 0)$, and this is the tangle in Figure 13.4.3(a) (cf. Chapter 9, Section 2), while in the case $\frac{u}{v} = \frac{1}{3}$, the corresponding tangle is $T(3, 0)$, and this is the tangle in Figure 13.4.3(b).

In a similar way, we may investigate the three other possibilities. For the case $(u, rv) = (9, -7)$,

$$\left\{\left(\frac{u}{v}, r\right)\right\} = \left\{\left(-\frac{9}{7}, 1\right), \left(\frac{9}{7}, -1\right), (-9, 7), (9, -7)\right\}.$$

As above, we may throw away the final two solutions and concentrate our attention on the first two possibilities. Since

$$\frac{9}{7} = 1 + \frac{1}{3 + \frac{1}{2}},$$

the corresponding tangle is $T(2, 3, 1)$, and with due consideration of the minus signs, these are the two tangles in Figure 13.4.3(c) and (d).

Finally, if S is an integer tangle, then we may show S does not satisfy (13.4.1)(2) and (3). The process is almost the same as above, but a touch more complicated. For example, suppose $S = T(s)$, where s is an integer.

We may now suppose the rational number corresponding to R is $\frac{u}{v}$ ($v > 1$). As above, if we again use Theorem 9.3.5, we shall obtain the following formula:

$$|vs + u| = 2. \qquad (13.4.4)$$

Exercise 13.4.3. Show that if S is an integer tangle, then

$$N(S + R + R) = N(R + (R + S)).$$

Since $R + S$ is a rational tangle, from Theorem 9.3.5 the determinant of $N(R + S) = N(S + R)$ is $|u + vs|$. Similarly, the determinant of $N(S + (R + R))$ is the absolute value of the numerator of

$$\frac{u}{v} + \frac{u}{v} + s = \frac{2uv + sv^2}{v^2},$$

which can be shown to be $|2uv + sv^2| = 5$. Hence, $v = 5$ and

$$|2u + sv| = 1. \qquad (13.4.5)$$

Exercise 13.4.4. Confirm the following are solutions to (13.4.4) and (13.4.5): $v = 5$ and

$$\{(u, s)\} = \left\{\left(-1, \frac{3}{5}\right), \left(1, -\frac{3}{5}\right), (-3, 1), (3, -1)\right\}.$$

Since the first two do not give integer solutions for s, we may ignore them. The final two, on the other hand, do not satisfy (13.4.1)(3) (show this by drawing the diagrams). Hence, we have proven the first of Theorem 13.4.1.

Exercise 13.4.5. Show, by drawing the relevant diagrams, the second part of Theorem 13.4.1.

To show, with the above tangles, that we are along the right lines for a correct model of the effect of the recombinase Tn3, we can confirm by drawing the relevant diagram that [13.4.1(5)] also holds.

Hence, from the above it should be possible to predict the result of further recombinations (in reality, this *is* possible).

By means of Theorem 13.4.1, we can determine $\{S, R\}$; however, the theorem does not shed any light on what E may be. It is not possible with certainty to be able to understand the structure of E. However, to a certain degree some of the structure can be seen from the following theorem:

Theorem 13.4.2.

Suppose there is a tangle E that satisfies the following two conditions:

(1) $N(T(-3, 0) + E) = C(1)$
(2) $N(T(-3, 0) + T(1)) = C(2)$ *with linking number* -1. $\qquad (13.4.6)$

If, in addition, E is a rational tangle, then E is a tangle of the form $T(2x, 3, 0)$, where x is an arbitrary integer.

Exercise 13.4.6. Confirm, by drawing the diagrams that for arbitrary n, $E = T(n, 3, 0)$ satisfies (13.4.6)(1).

If we add the second condition $(13.4.6)(2)$, then $T(-3,0)$ has an orientation induced on it. With regard to this orientation, if the site-specific recombination by the Tn3 resolvase is a direct repeat, then E must be of the form $T(2x,3,0)$.

In this chapter, we have seen via knot theoretical techniques that we may shed some light and elucidate the models for the recombination of a DNA molecule. At present, the extent knot theory may further help in the understanding of the mechanism of recombination of the DNA molecule is not clear. This area of research is still in its inchoate stages. Hopefully in the future this interaction will lead to some interesting, maybe coruscating, results.

Graph Theory Applied to Chemistry

In our discussions thus far we have considered a *graph* to be a figure, to put it naively, composed of dots and line segments (topologically this is called a 1-complex). To be more exact, less intuitive, and more mathematical, a graph is usually thought of in an abstract sense. Therefore, strictly speaking, a (finite) graph G is a pair of (finite) sets $\{V_G, E_G\}$ that fulfills an *incidence relation*. An element of V_G is then said to be a *vertex* of G, while an element of E_G is said to be an *edge* of G. The relation/condition mentioned above stipulates that an element, e, of E_G is *incident* to elements, say, a and b, of V_G (*nota bene*, the

condition does not require a and b to be distinct.) The two vertices a
and b are said to be endpoints of e. If it is the case that a = b, then e
is said to be a loop.

If there exist between two graphs, $G = \{V_G, E_G\}$ and $G' = \{V'_G, E'_G\}$, 1-1 correspondences $f_V : V_G \longrightarrow V'_G$ and $f_E : E_G \longrightarrow E'_G$
that satisfy condition (14.0.1) given below, then G and G' are said to
be *isomorphic*.

> If a vertex, a, of G is an endpoint of an edge, e, of G,
> then $f_V(a)$ is also an endpoint of $f_E(e)$. Conversely,
> if a vertex a' of G' is an endpoint of an edge, e',   (14.0.1)
> of G', then $f_V^{-1}(a')$ is also an endpoint of $f_E^{-1}(e')$.

Graph theory often relies on – as a way of illustrating its concepts and research results – *models* of graphs in space, i.e., the before-
mentioned 1-complexes. Frequently, the model of a graph and the abstract graph itself are perceived to be one and the same. However,
the model is a mere tool for presentation and expository purposes, and
should not really be confused with the abstract graph.

In a sense it is quite easy to come to the conclusion that an abstract
graph is some sort of generalization of a knot (or link); however, if we
are to be precise in our definition of a graph, this strictly is not the case.
For if we consider problems concerning graphs, the obvious problem, the
shape of the graph in space, is not a problem in the theory of (abstract)
graphs. This problem is dealt with separately in the theory of spatial
graphs, which *may* be thought of as a generalization of knots (or links);
we shall consider this in more detail in Section 2. For example, in
Figure 14.0.1 we have shown two models of the same graph; however,
as spatial graphs they are distinct.

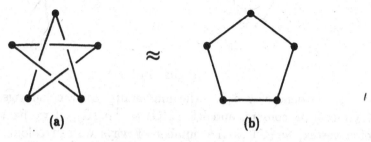

Figure 14.0.1

In this chapter we will firstly define an abstract graph invariant,
from which we may derive a spatial graph invariant. Since a spatial
graph *may* be thought of as a generalization of a knot (or link), then

this invariant is a generalization of a knot invariant. Having established this graph invariant, in the final section we shall look at the chiral properties of spatial graphs, which are of interest to chemists since it is possible to relate spatial graphs to the structure of molecules.

§1 An invariant of graphs: the chromatic polynomial

In the same way as for invariants of knots, a quantity that has the same value for two isomorphic graphs is said to be an invariant of graphs. Hence, no matter with what type of models we represent this graph, the value on the model will be invariant.

For example, the number of vertices, i.e., the number of edges of a graph, are the most obvious invariants. Besides such numerical invariants, it is possible also to construct polynomial invariants. As a typical example we shall instead take something called the *chromatic polynomial*.

Let G be a graph. We shall colour the vertices of the graph using a palette of n colours. The way to apply these colours is to paint two adjacent vertices (the vertices are the endpoints of an edge) with *distinct* colours. The above process is said to be *a (vertex) colouring of* G. Although we have on our palette n colours, it is not necessary to use all of them in the colouring process, Figure 14.1.1(a).

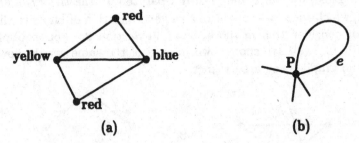

(a)   (b)

Figure 14.1.1

Let us denote by $\tilde{\rho}_n(G)$ the number of possible colourings of G with (at most) n colours, and let $\rho_n(G) = \frac{1}{n}\tilde{\rho}_n(G)$. If we fix the colour of a vertex, $\rho_n(G)$ is the number of ways we can colour the other vertices using n colours. Hence, this number can never be negative. From $\rho_n(G)$ for various n we can define, in the following manner, a polynomial (to be precise a power series) $\rho_G(t)$:

$$\rho_G(t) = \rho_1(G)t + \rho_2(G)t^2 + \cdots + \rho_n(G)t^n + \cdots \qquad (14.1.1)$$

The power series $\rho_G(t)$ is called the *chromatic polynomial* of G.

If G has a loop e [Figure 14.1.1(b)], then $\rho_n(G)$ becomes zero. In this case, since the end point P of the loop is self-adjacent, we cannot assign a colour to P. Let us calculate $\rho_n(G)$ for several examples.

Example 14.1.1. Suppose G is a graph consisting of only m vertices, i.e., G has no edges, Figure 14.1.2(a). Then each vertex can be coloured totally independently of the rest; hence, $\tilde{\rho}_n(G) = n^m$. Therefore,

$$\rho_G(t) = \sum_{n=1}^{\infty} n^{m-1} t^n.$$

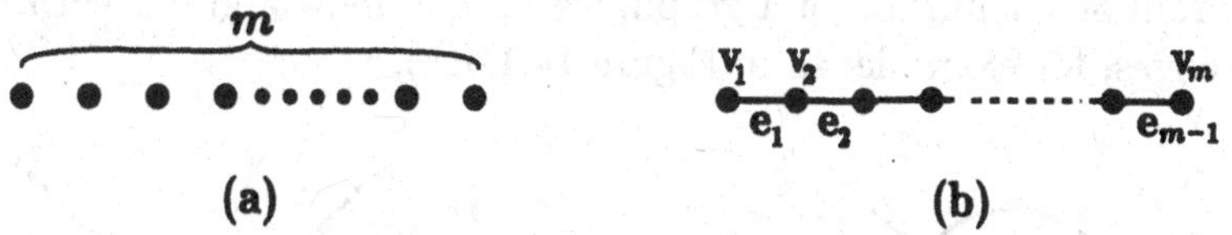

(a) **(b)**

Figure 14.1.2

Example 14.1.2. Suppose G is a graph consisting of m vertices, $\{v_1, v_2, \ldots, v_m\}$, and $m-1$ edges, $\{e_1, e_2, \ldots, e_{m-1}\}$, such that the endpoints of e_i are v_i and v_{i+1} $(i = 1, 2, \ldots, m-1)$, Figure 14.1.2(b). We may colour the vertex v_1 with n colours. The next vertex, v_2, since it cannot be coloured with the same colour as v_1, can be coloured with $n-1$ possible colours. Similarly, it is possible to colour v_3 with $n-1$ possible colours, and so on. This leads to

$$\tilde{\rho}_n(G) = n(n-1)^{m-1}.$$

Therefore,

$$\rho_G(t) = \sum_{n=1}^{\infty} (n-1)^{m-1} t^n.$$

Exercise 14.1.1. Calculate $\rho_G(t)$ when G is a polygon with m edges.

Let us denote by G_e the graph obtained from G by removing a single edge e, Figure 14.1.3(a). Similarly, let us denote by G/e the graph obtained from G by contracting e so that its two endpoints amalgamate, Figure 14.1.3(b). This latter process is called the *contraction* (with respect to e) of G. If e is a loop then $G_e = G/e$.

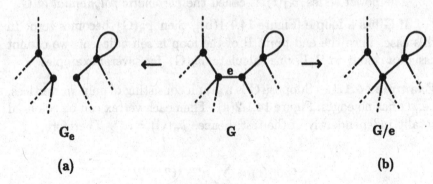

G_e G G/e

(a)   (b)

Figure 14.1.3

From our definition of a graph, we do not allow a graph with multiple edges, for example, as in Figure 14.1.4(a).

Figure 14.1.4

However, the contraction may produce multiple edges in G/e. Then by removing all but one of the edges, we can make G/e conform to our definition (see also Example 14.1.3).

Exercise 14.1.2. Show that the three graphs G, G_e, and G/e are related by the following equation:

$$\rho_n(G) = \rho_n(G_e) - \rho_n(G/e). \qquad (14.1.2)$$

Since $\rho_n(G_e)$ and $\rho_n(G/e)$ have at least one edge less than G, we can determine $\rho_n(G)$ by using (14.1.2) and mathematical induction. If G is a graph without any edges, then this is the graph described in Example 14.1.1, and so $\rho_n(G) = n^{m-1}$. This leads us to the next theorem.

Theorem 14.1.1.

The colouring number, $\rho_n(G)$, of a graph G coloured with (at most) n colours can be calculated by means of the following two formulae.

   (1) If G consists of m vertices, then

$$\rho_n(G) = n^{m-1}. \qquad (14.1.3)$$

(2) *If* e *is an edge of* G, *then*

$$\rho_n(G) = \rho_n(G_e) - \rho_n(G/e).$$

Example 14.1.3.

$$\rho_n\left(\triangle\right) = \rho_n\left(\angle\!\!\!\!\cdot\right) - \rho_n\left(\bigcirc\right) = \rho_n\left(\bullet\!\!-\!\!\bullet\!\!-\!\!\bullet\right) - \rho_n\left(\bullet\!\!-\!\!\bullet\right)$$

$$= (n-1)^2 - (n-1) = (n-1)(n-2).$$

A graph is said to be a *complete graph* if it has no loops and two distinct vertices are always the endpoints of only one edge. If K_m denotes a complete graph with m vertices, then K_m has $\frac{m(m-1)}{2}$ edges. In Figure 14.1.5 we have drawn models for complete graphs with $m \le 6$.

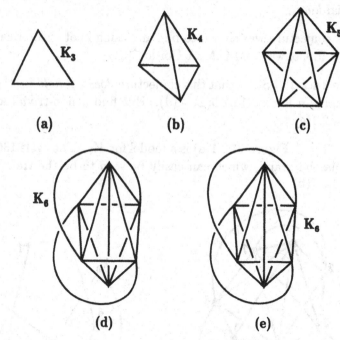

Figure 14.1.5

Exercise 14.1.3. Find a formula to calculate $\rho_n(K_m)$.

§2 Bing's conjecture and spatial graphs

We would like once again to underline the fact, mentioned in the previous section, that there is a clear distinction between a spatial graph as a model of an abstract graph and the abstract graph itself. Again, as noted previously and importantly, a spatial graph may be considered to be a generalization of a knot (or link). Along this vein, it is possible to open a seam for a theory of spatial graphs and then mine this seam. As a first step in an attempt to extract some information from this seam, we shall consider a conjecture due to R.H. Bing, whose main interest during his lifetime, however, was to find the right seam to quarry for a solution for the *Poincare conjecture*.

Bing's conjecture.
   Suppose that K_m is a complete graph with m vertices. If $m \geq 7$, then regardless of the spatial graph we use as a model for K_m, we can find within these spatial graphs a partial graph that represents a non-trivial knot.

   This conjecture was shown to be true, using knot theoretical techniques, by J. Conway and C.McA Gordon [CG].

Exercise 14.2.1.   Show that this conjecture *does not hold* for the complete graphs in Figure 14.1.5(a) ∼ (d). But find a non-trivial knot in Figure 14.2.1(e).

   For $m = 7$, Figure 14.2.1(a) is a model for K_7. The cycle 13642571 forms a partial graph, which can easily be seen to be the trefoil knot, Figure 14.2.1(b).

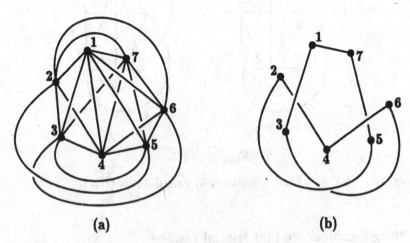

Figure 14.2.1

Exercise 14.2.2.   By drawing other spatial graph models for K_7, find other partial graphs that are not trivial knots.

From the above, spatial graph theory may be thought of as an extension of knot theory, rather than being a part of graph theory. So the research into spatial graphs may be tailored accordingly, i.e., with recourse to knot theory. However, if we are to consider spatial graphs as generalizations of knots (and links), we must include in our ruminations graphs of the type shown in Figure 14.2.2, i.e., ones that have closed curves with no vertices.

Figure 14.2.2

However, since a graph without vertices is an oxymoron, this type of spatial graph is not a model (in space) of a graph. Rather, it is better to think of a spatial graph as the *underlying space* of a 1-dimensional finite complex in $\mathbf{R}^3$. A vertex in this context is a point at which at least three arcs emerge. Also, it is possible that a spatial graph may have multiple edges. In this respect, we shall define equivalence of two spatial graphs.

Definition 14.2.1. If for two spatial graphs G_1 and G_2 there exists an auto-homeomorphism, φ, which preserves the orientation of $\mathbf{R}^3$ such that $\varphi(G_1) = G_2$, then G_1 and G_2 are said to be *equivalent* (or *equal*).

In the past, research into spatial graphs has been closely connected with the theory of surfaces embedded in $\mathbf{R}^3$. One such typical example is the study of the Kinoshita θ-curve. This graph is a spatial graph with two vertices, A and B, and three curves connected to them, and these curves do not mutually intersect each other.

Significantly, the two spatial graphs in Figure 14.2.3(a) and (b), with the vertices fixed in space, cannot be continuously deformed into each other. Therefore, as *spatial graphs* they are not "equivalent" (see also Definition 14.2.1).

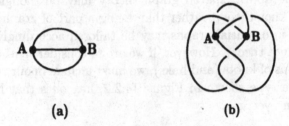

(a)   (b)

Figure 14.2.3

Graphs like the θ-curve, which have at each vertex exactly three incident edges, are called 3-regular graphs. The Jones polynomial, studied in Chapter 11, may be generalized so that it becomes an invariant of these graphs. However, we shall first take a quick look at a special set of spatial graphs, namely, plane graphs, which we previously encountered in Chapter 2, Section 3.

If we can place a spatial graph G on S^2, then G is said to be a *planar graph*. The graph that lies on S^2 is said to be a *plane graph*. With regard to a plane graph H, we can construct a graph $\tilde{H}$, called the *dual graph*, by means of the following procedure: Firstly, let us divide, by means of H, S^2 into a finite number of regions, $R_1, R_2, \ldots, R_m$, and from (within) each region R_i select a point $\tilde{v}_i$. These $\tilde{v}_i$ will now become the vertices of $\tilde{H}$. The edges of $\tilde{H}$ are polygonal arcs $\tilde{e}$ that are connected to the vertices of $\tilde{H}$ as described below.

If (and only if) two regions R_i and R_j have an edge, e, of H in common, then join $\tilde{v}_i$ and $\tilde{v}_j$ on S^2 by a simple polygonal line that intersects e at *only one*   (14.2.1) point. [In Figure 14.2.4(a) we have depicted these polygonal lines by dotted lines.]

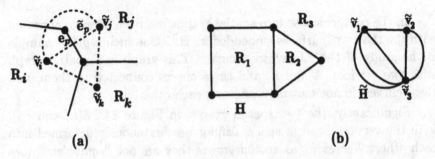

(a)   (b)

Figure 14.2.4

In Figure 14.2.4(b), we have drawn the dual graph $\tilde{H}$ of H. $\tilde{H}$ is uniquely determined from the plane graph H, which itself is a representation of the planar graph G; however, it is not uniquely determined from G.

Exercise 14.2.3. The two plane graphs H_1 and H_2 in Figure 14.2.5 are representations of a planar graph G. Draw the respective dual graphs $\tilde{H}_1$ and $\tilde{H}_2$.

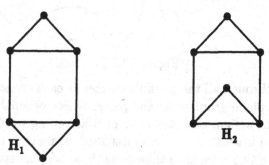

Figure 14.2.5

Exercise 14.2.4. Suppose G is a planar graph and H_1 and H_2 are plane graph representations of G. Show that if $\tilde{H}_1$ and $\tilde{H}_2$ are their respective dual graphs, then for all n

$$\rho_n(\tilde{H}_1) = \rho_n(\tilde{H}_2).$$

Further, show that this equality holds for the respective $\tilde{H}_1$ and $\tilde{H}_2$ in Exercise 14.2.3.

Now, as in the case of knots (or links), if we project the spatial graph G onto the plane, we can obtain a regular diagram, D, of G, Figure 14.2.6.

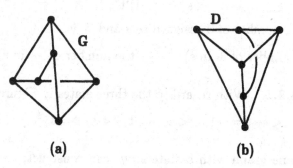

(a)   (b)

Figure 14.2.6

However, a vertex of G should not translate via the projection to a crossing point of D.

Suppose m is the number of crossing points of D. We obtain a three plane graph by replacing a crossing point (not a vertex) of the regular diagram D by one of the three local diagrams shown in Figure 14.2.7.

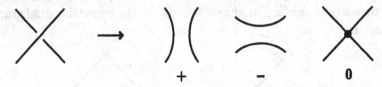

$$+ \qquad - \qquad 0$$

Figure 14.2.7

Running through all the possible choices at each crossing point, we can draw 3^m plane graphs. At the places where originally there was a crossing point assign, one of $+$, $-$, or 0 according to Figure 14.2.7. Each plane graph with $+$, $-$, or 0 assigned as above is called a *state* of D. [In Figure 14.2.8 we have shown the three possible states from the above operation on the regular diagram in Figure 14.2.6(b).]

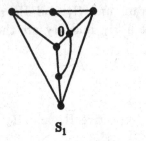  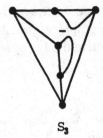

$$S_1 \qquad\qquad S_2 \qquad\qquad S_3$$

Figure 14.2.8

To each state, s, we can assign a monomial denoted by $< s >$,

$$< s > = (-1)^l t^k,$$

where l is the number of 0 assigned to s and k is

(the number of $+$ in s) $\quad - \quad$ (the number of $-$ in s).

Example 14.2.1. With regard to the three states in Figure 14.2.8,

$$< s_1 > = -1, \quad < s_2 > = t \quad < s_3 > = t^{-1}.$$

For a plane graph with a state s, we can construct, as described above, its dual graph, which we shall denote by $\tilde{s}$. In fact, $\tilde{s}$ is a genuine plane graph (without multiple edges). Then in conjunction with the colouring number, $\rho_n(\tilde{s})$, with n colours, we may define another polynomial, denoted by $< D >$, for the regular diagram D of G:

$$< D >= \sum_s < s > \rho_n(\tilde{s}),$$

where the sum on the right-hand side is taken over all the states s (3^m in total) on D. Performing the calculation, it is easy to see that $< D >$ is a polynomial in t and n; so if we replace n by $t + 2 + t^{-1}$, $< D >$ is transformed into a polynomial only in t, which we shall call the *polynomial* for D.

Theorem 14.2.1 [Y].

The polynomial for D, up to multiplication by a sign and t to some exponent, is an invariant of the 3-regular spatial graphs G. In other words, if D_1 and D_2 are, respectively, the regular diagrams of two equivalent 3-regular spatial graphs G_1 and G_2, then for some integer p,

$$< D_1 >= \pm t^p < D_2 > .$$

Therefore, the common value of the polynomial for G, ignoring the exponent of t and its sign, is an invariant of G and denoted by $\lambda_G(t)$. [This invariant is sometimes or also called the *Yamada* polynomial.]

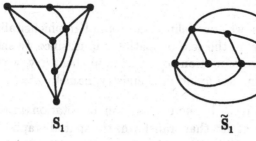

$$S_1 \qquad\qquad \tilde{S}_1$$

Figure 14.2.9

Example 14.2.1 (continued).

$$\lambda_G(t) =< s_1 > \rho_n(\tilde{s}_1)+ < s_2 > \rho_n(\tilde{s}_2)+ < s_3 > \rho_n(\tilde{s}_3)$$
$$= (-1)\rho_n(\tilde{s}_1) + t\rho_n(\tilde{s}_2) + t^{-1}\rho_n(\tilde{s}_3),$$

where $\tilde{s}_1$ is the dual graph for the state s_1, Figure 14.2.9.

Exercise 14.2.5. Using Example 14.2.1, show the following formulae for $\rho_n(\tilde{s}_i)$ hold, and hence determine $\lambda_G(t)$:

$$\rho_n(\tilde{s}_1) = (n-1)(n-2)(n^3 - 8n^2 + 23n - 23)$$
$$\rho_n(\tilde{s}_2) = (n-1)(n-2)(n-3)^2$$
$$\rho_n(\tilde{s}_3) = (n-1)(n-2)^3.$$

In the special case when G is a knot (or link), the invariant of G, $\lambda_G(t)$, agrees, up to a factor $\pm t^p$, with the 3rd degree Jones polynomial, $\widehat{J}_K^{(3)}$ (cf. Example 12.3.3). Therefore, $\lambda_G(t)$ may be considered as a generalization of the Jones polynomial.

Exercise 14.2.6. Show that if G^* is the mirror image of the spatial graph G, then

$$\lambda_{G^*}(t) = \lambda_G(t^{-1}).$$

Exercise 14.2.7. Calculate the polynomial of the two θ-curves in Figure 14.2.3, and show they are not equivalent.

Exercise 14.2.8. Calculate $\lambda_K(t)$ for the right-hand trefoil knot and compare it with its 3rd degree Jones polynomial, $\widehat{J}_K^{(3)}$.

§3 The chirality of spatial graphs

As for knots, we may define a concept of amphicheirality for spatial graphs. However, in the case of spatial graphs, since we shall discuss in this section their connection to concepts in chemistry, we shall use the terms more commonly found in chemistry, namely, *chiral, achiral*.

Definition 14.3.1. If there exists an orientation preserving auto-homeomorphism of R^3 that transforms the spatial graph G in R^3 to its mirror image G^*, then G is said to be topologically *achiral*. Otherwise, G is *chiral*. We can make this condition stronger, and say G is *rigidly achiral*, if we can rotate G, about some axis, to its mirror image G^*.

It is an immediate consequence that if G is rigidly achiral then it is also topologically achiral. However, the converse is not true.

Exercise 14.3.1. Show that a plane graph is rigidly achiral.

Exercise 14.3.2.   Show that the spatial graph in Figure 14.3.1 is topologically achiral but not rigidly achiral.

Figure 14.3.1

The question whether or not a spatial graph is achiral is of particular interest to chemists. For we may think of a vertex as an atom of some molecule. If in this molecule two atoms have a common bond, we may represent this by an edge connecting the two vertices that correspond to the atoms. In this way, the spatial graph we construct represents the structure (model) of a molecule. An attempt to create a molecule that has in its molecular structure a spatial graph was undertaken at the beginning of the 20th century. However, it was only in 1981 that the chemists H. Simmons and A. Paquette managed to synthesize a molecule with the graph in Figure 14.3.2 in its molecular structure.

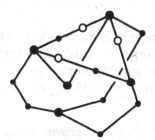

Figure 14.3.2

In Figure 14.3.2, the symbol "o" denotes an oxygen atom, and the other atoms are carbon atoms, but the hydrogen atoms have been omitted. Before this molecule was found, D.M. Walba had been successful in synthesizing a molecule whose molecular structure was a Möbius band M_3, Figure 14.3.3(a).

M_3, more generally M_n [Figure 14.3.3(b)], and its mirror image M_n^*, from the point of view of chemistry, do not mutually change, so they should not be equivalent, i.e., it is conjectured that M_n is chiral. In this section, we shall show that this conjecture can be solved by applying knot theoretical techniques.

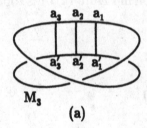

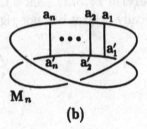

Figure 14.3.3

Theorem 14.3.1.
   The Möbius band M_n $(n \geq 4)$ is topologically <u>chiral</u>. Therefore, it cannot be rigidly <u>achiral</u>.

Theorem 14.3.2.
   If $n = 3$ there does not exist an (orientation preserving) auto-homeomorphism of $\mathbf{R}^3$ that transforms M_3 to M_3^* and

$$a_i a_i' \text{ to its mirror image } b_i b_i' \text{ for each } i = 1, 2, 3. \qquad (14.3.1)$$

Exercise 14.3.3. Show that if we remove condition (14.3.1), we can transform M_3 to M_3^*.

Exercise 14.3.4. Suppose $n \geq 4$, show that if there is an orientation preserving auto-homeomorphism, f, of $\mathbf{R}^3$, which transforms M_n to M_n^*, then there exists an orientation preserving auto-homeomorphism, g, of $\mathbf{R}^3$, which transforms M_n to M_n^* *and* transforms $a_i a_i'$ to $b_i b_i'$ for each $i = 1, 2, \ldots, n$.

   Since Theorem 14.3.1 can be proven using Theorem 14.3.2 and Exercise 14.3.4, we shall only prove Theorem 14.3.2 (cf. Exercise 14.3.5).

<u>Proof of Theorem 14.3.2.</u>
   Suppose that the auto-homeomorphism of Theorem 14.3.2 *exists* and denote it by h. Firstly, let us change M_3 to the form in Figure 14.3.4(a).
   Since the cycle $C = a_1 a_2 a_3 a_1' a_2' a_3' a_1$ is a trivial knot, the 2-fold cyclic covering space $\tilde{M}$ of S^3 branched along C is S^3 (cf. Chapter 8, Section 3, here we assume the graph is in S^3). In $\tilde{M}$ $(= S^3)$, the three semicircles $\alpha_i = a_i a_i'$ are extended to three circles $\tilde{\alpha}_i = \alpha_i \cup \alpha_i'$, as shown in Figure 14.3.4(b) (cf. Exercise 8.3.1). These three circles form the same link L as in Figure 4.5.6.

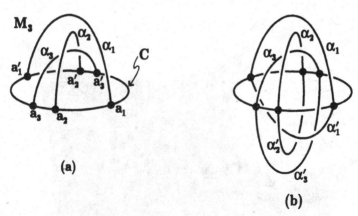

(a)

(b)

Figure 14.3.4

Similarly, we can consider the 2-fold cyclic covering space $\widetilde{M}^*$ of S^3 branched along the trivial knot $C^* = b_1 b_2 b_3 b_1' b_2' b_3' b_1$ in M_3^*, the mirror image of M_3. As above, the semicircles $\beta_i = b_i b_i'$ form circles $\tilde{\beta}_i = \beta_i \cup \beta_i'$ in $\widetilde{M}^*$ $(= S^3)$, Figure 14.3.5(b). These three circles form a link L^* that is the mirror image of L. Since L is chiral (Exercise 4.5.7), there does not exist an orientation preserving homeomorphism of $\mathbf{R}^3$ (and S^3) onto itself that maps L to L^*. It contradicts the existence of h, since h can be extended to an (orientation-preserving) homeomorphism from M to M^* that maps L to L^*.

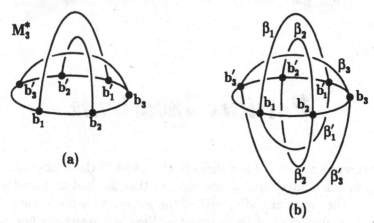

(a)

(b)

Figure 14.3.5

Exercise 14.3.5. Prove that for $n \geq 4$, M_n is topologically chiral.

Exercise 14.3.6. Calculate the graph invariant, $\lambda(t)$, for M_4, and hence show M_4 and M_4^* are not equivalent (cf. Exercise 14.2.6).

Vassiliev Invariants

Towards the end of the 1980s in the midst of the Jones revolution, V.A. Vassiliev introduced a new concept that has had profound significance in the immediate aftermath of the Jones revolution in knot theory [V]. The importance of these so-called Vassiliev invariants lies in that they may be used to study Jones-type invariants more systematically.

The other significance of the Vassiliev invariants to knot theorists is in the fact that they give a certain topological interpretation to the Jones-type invariants. But, to some dismay, this topological interpretation is not along classic lines. So, although the Vassiliev invariants add further insight to the Jones-type invariants, the question of how the Jones-type invariants relate to classic knot theory, for example, covering spaces, remains open.

We shall see in this chapter that the exact form of the Vassiliev invariants differs from anything we have discussed thus far in this book. The knot invariants, like the Alexander polynomial, associate a knot with some sort of mathematical quantity. A Vassiliev knot invariant, on the other hand, is an invariant that satisfies a set of conditions. In this sense, all the invariants introduced or redefined in Chapter 11 – the Jones polynomial, skein polynomial, and the Alexander polynomial – can all be shown to be Vassiliev invariants. However, not all the knot invariants we have defined in this book are Vassiliev invariants; we shall see later that the signature of a knot is not a Vassiliev invariant.

At the time of writing, the concept of Vassiliev invariants is still in a state of flux, with much of the research scattered in various journals and languages. We aim in this chapter to give an *elementary* introduction to this new concept, which will hopefully allow the reader to follow and maybe develop this concept.

§1 Singular knots

In the previous chapter we discussed spatial graphs, and in particular we defined an invariant of 3-regular graphs called the *Yamada polynomial*. The Vassiliev invariant is also connected to spatial graphs, and it can best be described as an invariant of certain *4-regular* spatial graphs called *singular knots*.

In its most basic understanding, a singular knot is a "knot" with self-intersections. More precisely, a singular knot, K, is (the image of) a (PL-)mapping $f : S^1 \to \mathbf{R}^3$, which is one-to-one except for a *finite* number points on S^1. Further, (1) if two points on S^1 have the same image, then $K = f(S^1)$ intersects itself at right angles at this common point, and (2) no three points on S^1 can have the same image.

Throughout this chapter, we shall call a point of self-intersection of a singular knot, K, a *vertex* of K. Some examples of singular knots are given in Figure 15.1.1 (a) ~ (f).

It is possible to extend the definition of a singular knot to the case of links; however, in the discussion that follows we will restrict ourselves to singular knots. Further, for convenience we shall not explicitly distinguish between singular knots and the usual kind of knots we have discussed in this book; however, if we refer just to a knot, then this will *never* mean a singular knot.

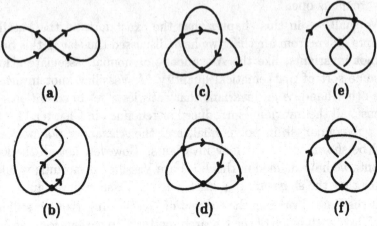

Figure 15.1.1

As mentioned above and as is now clear from our definition of a singular knot, we may consider a singular knot to be a 4-regular spatial graph. However, we need to refine what we mean by spatial graph theory, since a singular knot does not fall exactly into this category. Firstly, we may assign to a singular knot an orientation that is induced from the orientation on S^1; for spatial graphs, in general, there is no obvious way of assigning an orientation. Secondly, the notion of equivalence of singular knots is stronger than that for spatial graphs in general.

To be precise, let us associate with each vertex A of a singular knot K a small (closed) neighbourhood B_A and a plane E_A that contains $B_A \cap K$ – that part of K that is near A. Then a (plane) disk $F_A = E_A \cap B_A$ is called a *flat disk* associated with A. A collection of these flat disks, one for each vertex A, will be denoted by $\mathcal{F}(K)$.

Definition 15.1.1. Two singular knots K, K′ are said to be equivalent, denoted by $K \approx K'$, if there is an orientation-preserving auto-homeomorphism $\varphi : \mathbf{R}^3 \to \mathbf{R}^3$ that satisfies the following conditions:
  (1) $\varphi(K) = K'$;
  (2) There exist collections of flat disks $\mathcal{F}(K)$ and $\mathcal{F}(K')$ for K and K′, respectively, such that φ maps $\mathcal{F}(K)$ to $\mathcal{F}(K')$. (Such a transformation is called *flat*.)

Intuitively, K is equivalent to K′ if we can continuously deform K to K′ without causing any further self-intersections, and during the deformation the segments at the intersection points remain at right angles.

The second part of the above definition is important since therein lies the criterion that makes the equivalence of singular knots stronger than for spatial graphs in general.[15]

Example 15.1.1.

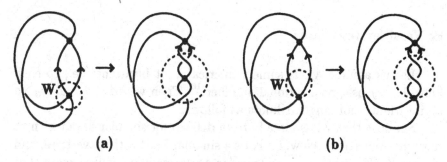

(a)   (b)

Figure 15.1.2

In each of Figure 15.1.2(a) and (b) we performed a vertical twist inside the ball W (keeping the outside fixed). In case (a) this produces two equivalent singular knots, but in case (b) they are not equivalent. However, cases (a) and (b) *are* equivalent as spatial graphs.

Example 15.1.2. The two singular knots (a) and (b) in Figure 15.1.1 are not equivalent, but again they are equivalent as spatial graphs. The same holds in the case of the two singular knots (c) and (d) of the same figure. (See also Exercise 15.3.3.)

In line with our previous discussion, we can, using singular diagrams, redefine the equivalence of singular knots.

Proposition 15.1.1.

Two singular knots, K and K′, are equivalent if and only if there exist singular diagrams D and D′, respectivley, which can be deformed into each other by applying a finite number of times (plane isotopy and)

(1) *The Reidemeister moves or their inverses except within a small neighbourhood of each vertex;*

(2) *The following operation, $\tilde{\Omega}$, near the vertices:*

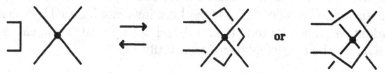

Figure 15.1.3

Exercise 15.1.1. (1) Show that the two singular knots (e), (f) in Figure 15.1.1 are equivalent.

(2) Show that the two singular knots in Figure 15.1.2(b) are equivalent to (c) and (d) in Figure 15.1.1.

§2 Vassiliev invariants

Let us assume v_0 is some numerical knot invariant; i.e., to each knot, K, v_0 assigns a rational number.[16] Then we may extend v_0 to an invariant v for singular knots as follows.

Suppose that v has already been defined for singular knots with at most $n-1$ vertices. Now, let K be a singular knot with n vertices, and further K, K_+ and K_- are (regular diagrams of) singular knots that are the same everywhere except at a neighbourhood of *one* vertex. In this neighbourhood, they differ only in the way shown in Figure 15.2.1.

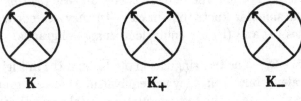

$$K \qquad\qquad K_+ \qquad\qquad K_-$$

Figure 15.2.1

Then we shall define

$$v(K) = v(K_+) - v(K_-) \tag{15.2.1}$$

Since K has n vertices and both K_+ and K_- have $n-1$ vertices, we may, by hypothesis, evaluate $v(K)$.

Exercise 15.2.1. Show that v defined by means of (15.2.1) is in fact an invariant for singular knots. In other words, if K and K' are equivalent singular knots, then $v(K) = v(K')$.

The above singular knot invariant, v, depends implicitly on the knot invariant v_0 from which we start the inductive process. Therefore, this invariant v will be called the singular knot invariant induced by the knot invariant v_0. The relation (15.2.1), which we may call a *singular skein relation*, can be rewritten diagrammatically as

$$v\left(\text{\ding{56}}\right) = v\left(\text{\ding{56}}\right) - v\left(\text{\ding{56}}\right). \tag{15.2.1a}$$

As might be expected from our previous discussions, we shall call the diagrams in Figure 15.2.1 singular skein diagrams. Further, we shall call the operation that replaces one diagram in Figure 15.2.1 by the other two skein diagrams a singular skein operation. When we apply the operation that replaces K_+ (or K_-) by K_- (or K_+) and K, it can be said that we are applying an unknotting operation that results in a new vertex.

Definition 15.2.1. A singular knot invariant v [satisfying (15.2.1)] is called a *Vassiliev invariant of order (at most)* m (or can also be said to be of *finite type*) if for any singular knot with $m + 1$ vertices,

$$v(K) = 0. \tag{15.2.2}$$

In particular, if v is of order at most m but not of order $m - 1$, i.e., there exists a singular knot with exactly m vertices for which v is non-zero, then v is called a Vassiliev invariant of order (exactly) m.

Exercise 15.2.2. Show that if v is a Vassiliev invariant of order at most m, then v vanishes for any singular knot with *at least* $m + 1$ vertices.

As can be seen from the above definitions, a Vassiliev invariant is essentially different from our previous knot invariants, for example, the Jones polynomial, the Alexander polynomial, and the bridge number. These former knot invariants associate some sort of mathematical quantity to each knot. Rather, a Vassiliev invariant is a singular knot invariant that satisfies (15.2.1) and (15.2.2). Therefore, there are (infinitely) many Vassiliev invariants, and it seems more natural to call them invariants of Vassiliev *type*.

Obviously there are many knot invariants that induce Vassiliev invariants; however, not *all* knot invariants induce Vassiliev invariants, for a more detailed discussion, see Section 3. Before we give some examples of the knot invariants that induce Vassiliev invariants, we will prove some propositions that follow easily from the definition of a Vassiliev invariant.

Proposition 15.2.1.

Let v be a Vassiliev invariant. If a singular knot K has a "loop"

, *then*

$$v(K) = 0.$$

<u>Proof</u>

If we apply (15.2.1) to the vertex that forms part of the loop, then we immediately write down the following equation:

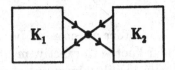

$$v(K) = v\left(\bigcirc\!\!\!\!\infty\right) = v\left(\bigcirc\!\!\!\!\infty\right) - v\left(\bigcirc\!\!\!\!\infty\right) = 0.$$

∎

Exercise 15.2.3. Show that if a singular knot K has a vertex as in Figure 15.2.2, then $v(K) = 0$.

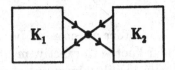

Figure 15.2.2

Proposition 15.2.2.

Let v be a Vassiliev invariant of order 0. Then for any non-singular knot K,

$$v(K) = v\left(\bigcirc\right).$$

Therefore, there is essentially only one Vassiliev invariant of order zero.

<u>Proof</u>

Since v is of order 0, it follows that $v\left(\times\!\!\!\!\bullet\right) = 0$, and hence from the singular skein relation $v\left(\times\right) = v\left(\times\right)$. This implies that if we apply an unknotting operation at any crossing point of K, then the value of v remains constant. Since a knot can be deformed to a trivial knot by applying several unknotting operations (cf. Proposition 4.4.1), it follows that $v(K) = v\left(\bigcirc\right)$. Therefore, v is a constant for any non-singular knot. ∎

Proposition 15.2.3.

There is no Vassiliev invariant of order (exactly) one.

<u>Proof</u>

Let us assume that there exists a Vassiliev invariant of order one, say, v. Then, by assumption, v is zero for any singular knot with two or more vertices. But there does exist a singular knot K_1 with a single vertex for which $v(K_1) \neq 0$. We can apply unknotting operations to K_1 which in turn increases the number of vertices, and thus K_1 can be deformed to the singular knot $\widehat{K}$ in Figure 15.1.1(a).

Since v vanishes for any singular knot with two or more vertices, it follows from (15.2.1) and Proposition 15.2.1 that $v(K_1) = v(\widehat{K}) = 0$, see Figure 15.1.3. However, this contradicts our original hypothesis. ∎

Let K be a singular knot with $m+1$ vertices, and number these vertices $1, 2, \ldots m+1$. Now let us apply a singular skein operation at each vertex to eliminate these vertices. This process will create 2^{m+1} knots $\{K_{\epsilon_1, \epsilon_2, \ldots, \epsilon_{m+1}}\}$ from K, where $\epsilon_1, \epsilon_2, \ldots, \epsilon_{m+1}$, are either $+$ or $-$, and $K_{\epsilon_1, \ldots, \epsilon_{m+1}}$ denotes the knot obtained from K by replacing the vertex k $(k = 1, 2, \ldots, m+1)$ by a positive crossing point if $\epsilon_k = +$, or by a negative crossing point if $\epsilon_k = -$. By repeatedly applying the singular skein relation, we obtain the following formula:

$$v(K) = \sum_{\epsilon_1, \epsilon_2, \ldots, \epsilon_{m+1}} (-1)^l v(K_{\epsilon_1, \epsilon_2, \ldots, \epsilon_{m+1}}), \qquad (15.2.3)$$

where the summation is taken over all the set of 2^{m+1} elements $\epsilon_1, \epsilon_2, \ldots, \epsilon_{m+1}$, and l is the number of $-$ in the sequence $\epsilon_1, \ldots, \epsilon_{m+1}$. Using (15.2.3) we may replace (15.2.2) by a more convenient form (15.2.4) of the following proposition.

Proposition 15.2.4.

A singular knot invariant (that satisfies (15.2.1)) is a Vassiliev invariant of order at most m *if and only if for any singular knot with* $m+1$ *vertices,*

$$\sum_{\epsilon_1, \epsilon_2, \ldots, \epsilon_{m+1}} (-1)^l v(K_{\epsilon_1, \epsilon_2, \ldots, \epsilon_{m+1}}) = 0, \qquad (15.2.4)$$

where the summation is taken over the same set as in (15.2.3).

To evaluate a Vassiliev invariant, it is as in the case of the skein polynomials more convenient to draw *singular skein tree* diagrams (cf. Example 6.2.1).

Example 15.2.1. In Figure 15.2.3 we have drawn the singular tree diagram to evaluate a Vassiliev invariant for a singular knot with two vertices.

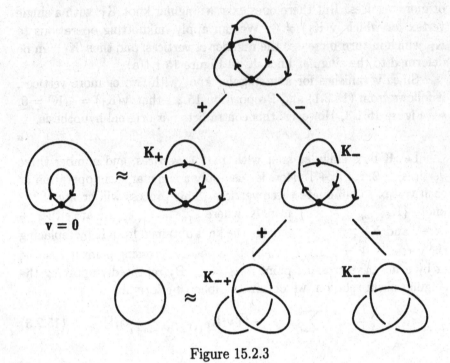

Figure 15.2.3

Therefore,

$$v(K) = -v(K_{-+}) + v(K_{--}) = -v\left(\bigcirc\right) + v(K_{--}).$$

Exercise 15.2.4. Draw singular skein diagrams for the singular knots in Figure 15.1.1.

§3 Some examples of Vassiliev invariants

In this section we shall give several concrete examples of Vassiliev invariants. We shall also show that there exist some knot invariants that are not of finite type, and hence are not Vassiliev invariants.

Proposition 15.3.1.

Let K be a knot and

$$\nabla_K(t) = 1 + a_2 z^2 + a_4 z^4 + \ldots + a_{2m} z^{2m} + \ldots$$

be the Alexander-Conway polynomial of K. Then $\nabla_{2m}(K) = a_{2m}$ $(m = 1, 2, \ldots)$ induces a Vassiliev invariant of type exactly $2m$.

Proof

It is sufficient to show that the following two conditions hold:

(1) For a singular knot with $2m + 1$ vertices, K, $\nabla_{2m}(K) = 0$;
(2) There is a singular knot with $2m$ vertices, K_0, such that $\nabla_{2m}(K_0) \neq 0$.

By our work in the previous section, we may rewrite (1) as

(1)' $\sum (-1)^l \nabla_{2m}(K_{\epsilon_1, \epsilon_2, \dots, \epsilon_{2m+1}}) = 0$.

Our aim will be to prove (1)' rather than (1). In this regard, let us denote by $(K_{\epsilon_1, \epsilon_2, \dots, \epsilon_i, 0, \dots, 0})$ the knot obtained from $K_{\epsilon_1, \epsilon_2, \dots, \epsilon_{2m+1}}$ by replacing the last $2m + 1 - i$ crossing points by ⦶. We are now in a position to apply the skein relation (6.2.1) that defines the Conway polynomial, and so obtain the following formulae:

For any $\epsilon_1, \epsilon_2, \dots, \epsilon_{2m+1}$,

$$\nabla_{K_{\epsilon_1, \epsilon_2, \dots, \epsilon_{2m}, +}}(z) - \nabla_{K_{\epsilon_1, \epsilon_2, \dots, \epsilon_{2m}, -}}(z) = z\nabla_{K_{\epsilon_1, \epsilon_2, \dots, \epsilon_{2m}, 0}}(z).$$

If we apply the skein relation once again, but this time to $K_{\epsilon_1, \epsilon_2, \dots, \epsilon_{2m}, 0}$, we obtain

$$\nabla_{K_{\epsilon_1, \dots, \epsilon_{2m-1}, +, 0}}(z) - \nabla_{K_{\epsilon_1, \dots, \epsilon_{2m-1}, -, 0}}(z) = z\nabla_{K_{\epsilon_1, \dots, \epsilon_{2m-1}, 0, 0}}(z).$$

We keep applying the skein relation until we reach the following equation:

$$\nabla_{K_{+, 0, \dots, 0}}(z) - \nabla_{K_{-, 0, \dots, 0}}(z) = z\nabla_{K_{0, 0, \dots, 0}}(z).$$

Collecting these terms together, we finally obtain

$$\sum (-1)^l \nabla_{K_{\epsilon_1, \epsilon_2, \dots, \epsilon_{2m+1}}}(z) = z^{2m+1}\nabla_{K_{0, 0, \dots, 0}}(z). \qquad (15.3.1)$$

Since the right-hand side of (15.3.1) does not contain the term z^{2m}, $\nabla_{2m}(K) = 0$.

Now turning our attention to the proof of (2), let us consider the singular knot $K[p, q]$, depicted in Figure 15.3.1, where p is the number of vertices and $|q|$ is the number of crossing points.

K[5,3]   **K[5,-3]**

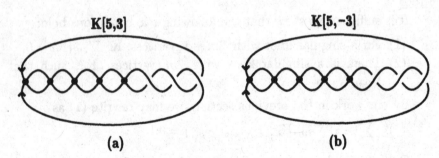

(a)   (b)

Figure 15.3.1

Let us look at the specific singular knot $K[2m,1]$ and compute $\nabla_{2m}(K[2m,1])$, the required formula is given in (15.3.2):

$$\nabla_{2m}(K[2m,1]) = \sum(-1)^l \nabla_{2m}(K[2m,1]_{\epsilon_1,\epsilon_2,\ldots,\epsilon_{2m}}). \qquad (15.3.2)$$

If we refer back to our discussion on torus knots and their generic form (Chapter 7), then it is easy to see that $K[2m,1]_{\epsilon_1,\epsilon_2,\ldots,\epsilon_{2m}}$ is a torus knot of type $(2m-2l+1,2)$, where l is the number of " $-$ " in the sequence $\epsilon_1, \epsilon_2, \ldots, \epsilon_{2m}$. Therefore, $\Delta(t) = 1$ if $l = m$ *or* $m+1$ and

$$\Delta(t) = t^{-(m-l)} + \ldots + t^{m-l} = 1 + \ldots + z^{2(m-l)} \qquad \text{if } 0 \le l < m,$$

$$\Delta(t) = t^{-(l-m-1)} + \ldots + t^{l-m-1} = 1 + \ldots + z^{2(l-m-1)} \quad \text{if } m+1 < l.$$

Therefore, $\nabla_{2m}(K[2m,1]_{\epsilon_1,\epsilon_2,\ldots,\epsilon_{2m}}) = 1$ if each ϵ_i is $+$ and zero otherwise. Hence, we may deduce that $\nabla_{2m}(K[2m,1]) = 1$.

∎

The above proposition shows that there exists a Vassiliev invariant induced from the Alexander-Conway polynomial, ∇_{2m}. The reader may ask, Can the same be said of the Jones polynomial? The next theorem answers this question in the affirmative.

Theorem 15.3.2 [BL].
Let $V_K(t)$ be the Jones polynomial of a knot K. Let $V_K(q)$ be the infinite series obtained from $V_K(t)$ by substituting e^q $(= 1 + q + \frac{q^2}{2!} + \ldots = \sum_{n=0}^{\infty} \frac{q^n}{n!}$ for t. So we may write

$$V_K(q) = b_0 + b_1 q + b_2 q^2 + \ldots .$$

Then $J_m(K) = b_m$ is a Vassiliev invariant induced by the Jones polynomial of order (at most) m.

As might be expected, it is also possible to define a Vassiliev invariant, $J_m^{(N)}$, induced from the Nth $(N \geq 2)$ degree Jones polynomial, $\widehat{J}_K^{(N)}(t)$, using the same substitution (see Example 12.3.3). However, we shall here discuss another way of obtaining a Vassiliev invariant from the Jones polynomial.

Proposition 15.3.3.
Let $\phi_K(t)$ be the Taylor expansion of the Jones polynomial of a knot K at $t = 1$, which we may write as

$$\phi_K(t) = c_0 + c(t - 1) + c_2(t - 1)^2 + \ldots + c_m(t - 1)^m + \ldots,$$

where if we recall some first-year calculus, $c_m = \frac{1}{m!}\left[\frac{d^m V_K(t)}{dt^m}\right]_{t=1}$.
Then $\phi_K(t) = c_m$ is a Vassiliev invariant of order (at most) m.

Exercise 15.3.1. (1) Prove Proposition 15.3.3.
(2) If K is the right-hand trefoil knot, show that
  (i) $\nabla_2(K) = 1$ and $\nabla_i(K) = 0$ for $i \geq 3$;
  (ii) $J_2(K) = -3$ and $J_3(K) = -6$;
  (iii) Evaluate $J_2^{(3)}(K)$, $J_3^{(3)}(K)$ (see Example 12.3.3), and $\phi_i(K)$
for $i \geq 2$.
(3) For the figure 8 knot, evaluate J_i for $i = 2, 3, 4$, ∇_i, and ϕ_i
for $i \geq 2$.
(4) For $K = K[4, 1]$, show that $J_2(K) = J_3(K) = 0$, and evaluate
$J_4(K)$.

Exercise 15.3.2. Show that if K^* is the mirror image of a knot K, then $J_m(K^*) = (-1)^m J_m(K)$ for all $m \geq 2$.

Exercise 15.3.3. (1) Show, by evaluating J_2 or ∇_2, that the two singular knots in Figure 15.1.1(a), (b) are not equivalent.
(2) Show, by evaluating J_3, that the two singular knots in Figure 15.1.1(c), (d) are not equivalent, and hence the two singular knots in Figure 15.1.2(b) are not equivalent (see Exercise 15.1.2).

Remark 15.3.1. It is possible, via the substitution $t = e^q$, to induce Vassiliev invariants from the skein polynomials and the Kauffman polynomial in the same way as in Theorem 15.3.2. However, since these polynomials have two variables, they need to reduced to 1-variable polynomials. In the case of the skein polynomial, $P_K(v, z)$, the polynomial $\widehat{J}_K(t)$ given in Example 12.3.4 is such a 1-variable specialization of $P_K(v, z)$. Not surprisingly, $\hat{J}_K(t)$ induces a Vassiliev invariant, for more details see Birman and Lin [BL].

We have seen that it is relatively easy to build up a variety of Vassiliev invariants induced from the knot invariants in Chapter 11. However, when we start to look at what we may call the "classic" geometric invariants rather than the "algebraic," then the stream of Vassiliev invariants dries up. Let us look a bit more closely at why this is the case.

Proposition 15.3.4.
The singular knot invariant, v, defined by the signature of a knot K is not of finite type; hence it is not a Vassiliev invariant.

Proof
Let us assume that v is of order at most $n \ (\geq 2)$.
We intend to show that $v(K[n+1, n]) \neq 0$, or equivalently,

$$\sum (-1)^l \sigma(K[n+1, n]_{\epsilon_1, \epsilon_2, \ldots, \epsilon_{n+1}}) \neq 0. \tag{15.3.3}$$

Now, since $K[n+1, n]_{\epsilon_1, \epsilon_2, \ldots, \epsilon_{n+1}}$ is a torus knot of type $(2n - 2l + 1, 2)$, its signature, see Theorem 7.5.1(IV), is

(i) If $l \neq n, n+1$, $\sigma(K[n+1, n]_{\epsilon_1, \epsilon_2, \ldots, \epsilon_{n+1}}) = -(2n - 2l)$;

(ii) If $l = n$ or $n+1$, $\sigma(K[n+1, n]_{\epsilon_1, \epsilon_2, \ldots, \epsilon_{n+1}}) = 0$.

It is quite easy to see that the number of knots with l negative signs in the sequence $\epsilon_1, \epsilon_2, \ldots, \epsilon_{n+1}$, is $\binom{n+1}{l}$. So the left-hand side of (15.3.3) is

$$\sum (-1)^l \sigma(K[n+1, n]_{\epsilon_1, \epsilon_2, \ldots, \epsilon_{n+1}}) = \sum_{l=0}^{n-1} (-1)^{l+1} \binom{n+1}{l} (2n - 2l),$$

which can be shown to be non-zero, see Exercise 15.3.4. This now contradicts the assumption that v is of order at most n.

■

Exercise 15.3.4. Show that if $n \geq 1$,

$$\sum_{l=0}^{n-1} (-1)^l \binom{n+1}{l} (n - l) = (-1)^{n+1}.$$

If we look closely at the proof of the previous proposition, we can easily prove the following proposition:

Proposition 15.3.5.

None of the following classical geometric invariants: the minimal crossing number $c(K)$, the unknotting number $u(K)$, the bridge number $br(K)$, the braid index $b(K)$ and the genus $g(K)$ of a knot, K is of finite type; hence, they cannot be Vassiliev invariants.

Exercise 15.3.5. By considering the singular knot $K[n+1, n]$, prove Proposition 15.3.5.

§4 Chord diagrams

Besides the equality in Proposition 15.2.1 (and Exercise 15.2.3), any Vassiliev invariant, v, satisfies a further important "linear" equation called the 4-term formula.

Theorem 15.4.1 (4-term formula).

The following equality holds for any Vassiliev invariant, v,

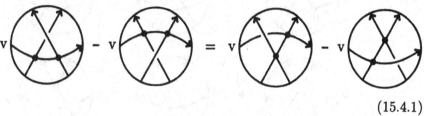

$$(15.4.1)$$

Figure 15.4.1

where the 4 singular knots are identical outside the circles.

The above "graphical" formula may be interpreted as the equivalent of a "Reidemeister move Ω_3" for singular knots. However, the keen reader may notice that there is in fact a slight difference between the over-crossing and under-crossing at each crossing point.

Proof

Let us consider the two equivalent singular knots K and K'. K and K' are identical, except inside the circles shown in Figure 15.4.2.

        (a)   **(b)**

Figure 15.4.2

If we apply singular skein operations at the crossing points a and b in the circles, Figure 15.4.3, then we obtain the following equalities:

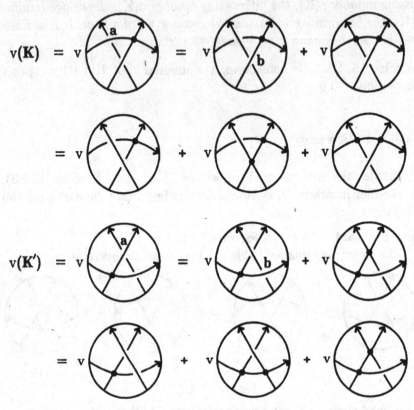

Figure 15.4.3

The first terms of the final expansions of $v(K)$ and $v(K')$ are equal; this follows because they are the v-value of equivalent singular knots. So, if we rearrange the remaining terms we obtain the desired formula (15.4.1).

∎

Exercise 15.4.1. Show that for any Vassiliev invariant, v, the following formula holds:

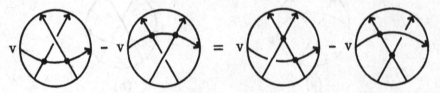

Figure 15.4.4

Due to Proposition 15.2.2, the Vassiliev invariant, v_0, of order 0 is essentially "unique." Further, we have shown in Proposition 15.2.3 that there is no Vassiliev invariant of order 1. Using the same ideas, we may determine the Vassiliev invariant of order (exactly) 2.

To this end, let us apply unknotting operations to a singular knot K, which will cause vertices to be added to the subsequent singular knots and also a trivial knot to be formed. Since v_2 vanishes for any singular knot with more than two vertices, we can write down the following equality:

$$v_2(K) = av_2\left(\bigcirc\right) + bv_2\left(\text{⊂⊃⊂⊃}\right) + cv_2\left(\text{⊂⊃⊂⊃⊂⊃}\right) + dv_2\left(\text{⊛}\right),$$
$$(15.4.2)$$

where a, b, c and d are integers. (See also Example 15.4.1 below.)

Example 15.4.1. Suppose K is the right-hand trefoil knot. To derive an expression for $v_2(K)$ in the form of (15.4.2), we make use of the singular skein diagram, see Figure 15.4.5.

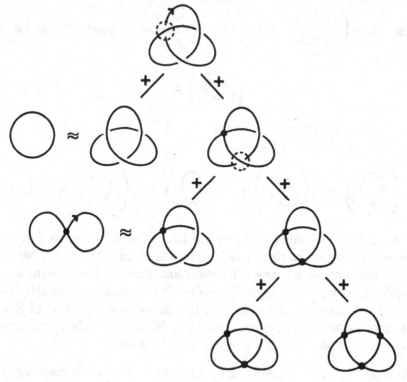

Figure 15.4.5

Since $v_2\left(\vcenter{\hbox{⊖}}\right) = 0$, we can write

$$v_2(K) = v_2\left(\bigcirc\right) + v_2\left(\infty\right) + v_2\left(\vcenter{\hbox{⊖}}\right).$$

Now, since $v_2\left(\infty\right) = 0$ and $v_2\left(\text{∞∞}\right) = 0$, v_2 is essentially determined by $v_2\left(\bigcirc\right)$ and $v_2\left(\vcenter{\hbox{⊖}}\right)$. However, analogously to $\nabla_2\left(\bigcirc\right)$ and $J_2\left(\bigcirc\right)$, it is natural to assign for any Vassiliev invariant of order m $(m \geq 2)$,

$$v_m\left(\bigcirc\right) = 0. \tag{15.4.3}$$

Therefore, it follows that $v_2(K)$ is completely determined by the value of $v_2\left(\vcenter{\hbox{⊖}}\right)$. Hence, v_2 is essentially *unique*.[17] So let us assign

$$v_2\left(\vcenter{\hbox{⊖}}\right) = 1. \tag{15.4.4}$$

(We may assign any non-zero number to it.)

Then the following equality

$$v_2\left(\vcenter{\hbox{⊖}}\right) = v_2\left(\vcenter{\hbox{⊖}}\right) + v_2\left(\vcenter{\hbox{⊖}}\right) = v_2\left(\vcenter{\hbox{⊖}}\right) = 1 \tag{15.4.5}$$

shows that the v_2-value of a singular knot K with 2 vertices does not depend on the sign of the crossing points in K. In other words, we can ignore the difference between an over- and under-crossing point for a singular knot with 2 vertices. Therefore, if we consider a singular knot K as a mapping $f : S^1 \to \mathbf{R}^3$, in the case of the v_2-value of K all that we need to concern ourselves with is, Which two points of the four points a, b, c, and d on S^1 have the same image?

Example 15.4.2. Consider a singular knot, K, as the mapping $f : S^1 \to \mathbf{R}^3$.

<u>Case 1</u> $f(a) = f(b)$ and $f(c) = f(d)$.

Although K and the singular knot in Figure 15.4.6(a) may not be equivalent, they have the same v_2-value.

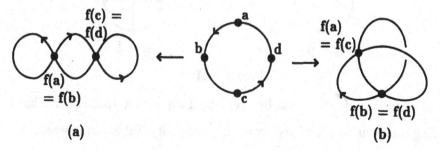

Figure 15.4.6

<u>Case 2</u> $f(a) = f(c)$ and $f(b) = f(d)$.

In this case, K and the singular knot in Figure 15.4.6(b) have the same v_2-value, even though they may not be equivalent, this follows from (15.4.5).

The remaining case, i.e., $f(a) = f(d)$ and $f(b) = f(c)$, can be shown to reduce to Case 1.

To emphasize the above connection between the four points, we shall assign diagram (a) in Figure 15.4.7 to Case 1 and diagram (b) of the same figure to Case 2. [As noted above, we may ignore the under- and over-crossing information of the intersection of the arcs in Figure 15.4.7(b).]

Figure 15.4.7

The diagrams in Figure 15.4.7 are usually called *chord diagrams*. Once we assign v_2-values to the two chord diagrams in Figure 15.4.7, it is possible to evaluate v_2 for any singular knot. The table of chord diagrams with their v_2-values is called the *Actuality Table*.

Actuality Table for v_2

$$V_2 = 0 \qquad\qquad V_2 = 1$$

Figure 15.4.8

Since $v_2\left(\text{⊘}\right)$ must be zero by Proposition 15.2.1, this chord diagram is usually omitted from the Actuality Table. In essence, the choice of value for $v_2\left(\oplus\right)$ is arbitrary as long as we do not assign zero to it.

It is possible to keep on building Actuality Tables *ad infinitum*; however, we shall only consider one more case – Vassiliev invariants of order 3, v_3.

In a similar way to (15.4.2), we can write for any singular knot, K,

$$v_3(K) = av_3\left(\text{⊖}\right) + bv_3\left(\text{⊗}\right) + cv_3\left(\text{⊖}\right), \qquad (15.4.6)$$

where a, b, and c are integers. In the above formula we have omitted those singular knots, for example, $\bigcirc$ and ∞, whose v_3-value, by (15.4.3) or Proposition 15.2.1, must be zero.

Exercise 15.4.2. Determine the values of a, b, and c in (15.4.6) for the right-hand trefoil knot and the figure 8 knot.

Therefore, to determine $v_3(K)$ it is sufficient to assign values to the singular knots in (15.4.6).

Caveat lector, the value of $v_3\left(\text{⊖}\right)$ in the Actuality Table need not necessarily be the same value as $v_2\left(\text{⊖}\right) = v_2\left(\oplus\right)$. The reason is that the v_3-value of a singular knot, K, with 2 vertices *may* depend on the *sign* of the crossing points of K. In fact, from (15.2.1),

$$v_3\left(\text{⊖}\right) = v_3\left(\text{⊖}\right) + v_3\left(\text{⊖}\right).$$

In comparison to v_2, there are 3 chord diagrams, shown below, for v_3.

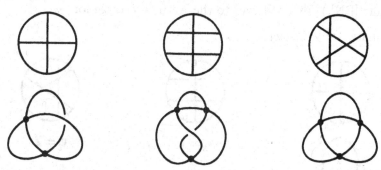

To construct the Actuality Table for v_3, we must assign a v_3-value to each of these three chord diagrams.

However, in contrast to the previous case, this is not so simple a matter. The sticking point is that we may no longer assign arbitrary values to the chord diagrams. For example, $v_3\left(\bigoplus\right)$ and $v_3\left(\bigotimes\right)$ are not independent. There is, in fact, an equation, see Figure 15.4.9, that involves the both of them.

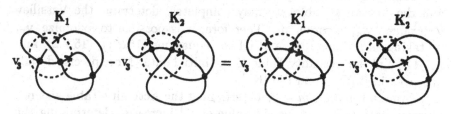

Figure 15.4.9

On close inspection, the four configurations inside the circles of the above figure can be seen to satisfy the 4-term formula of Theorem 15.4.1. It follows immediately from Proposition 15.2.1 that $v_3(K_2) = 0$.

Exercise 15.4.3 Show that $v_3(K_1) = v_3(K_2')$. (*Caveat lector*, K_1 and K_2' are not necessarily equivalent.)

The above exercise allows us to write down the following equality:

$$2v_3\left(\bigoplus\right) = v_3\left(\bigotimes\right).$$

In this way, we must find all the equations involving these chord diagrams and then assign v_3-values to these chord diagrams so that these equations are satisfied. (The situation is not as bad as it seems.

It is known that any equality that holds for these chord diagrams of any order is obtained "only" from the 4-term formula and variations on this formula [BN].) This will lead to the Actuality Table for v_3.

Actuality Table for v_3

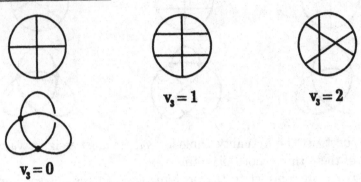

$$v_3 = 1 \qquad v_3 = 2$$

$$v_3 = 0$$

Figure 15.4.10

Since there is only one arbitrary choice for either $v_3\left(\bigoplus\right)$ or $v_3\left(\bigotimes\right)$, a Vassiliev invariant of order 3 is essentially unique.

Following the above procedure, i.e., by means of (15.2.1), (15.2.2), and the Actuality Table, we may completely determine the Vassiliev invariants, v_m, of order m. Therefore, it is possible to axiomize v_m by taking as Axioms I, II, and III the equations (15.2.1), (15.2.2), and (15.4.3) and the Actuality Table as the initial data. More precise details are given in Birman and Lin [BL].

Even in the case $m = 3$, determining the Actuality Table was not an easy matter. In fact, as the value of m increases, determining the Actuality Table requires tremendous computing power. For example, for $m = 8$ it is necessary to solve more than 300,000 linear equations with more than 40,000 unknowns! For the adventurous reader, further details may be found in Bar-Natan [BN].

Exercise 15.4.4. (1) Show that for the right-hand trefoil knot, K, and its mirror image, K*,

$$v_2(K) = 1 \text{ and } v_3(K) = 2; \quad v_2(K^*) = 1 \text{ and } v_3(K^*) = 0.$$

(2) Show $v_2(K) = -1$ and $v_3(K) = -1$ for the figure 8 knot, K.

Exercise 15.4.5. Show that for any knot K, $\nabla_2(K) = v_2(K)$. Is it possible to express the Vassiliev invariants J_2 and J_3 in terms of v_2 or v_3?

Exercise 15.4.6. (1) Suppose that a chord diagram D has a simple arc that is disjoint from the other arcs in D. Show that for any Vassiliev invariant, v, v(D) = 0.

(2) Find the chord diagrams for a Vassiliev invariant of order 4 whose v_4-value is not automatically zero by Proposition 15.2.1. (Hint: It is known there are 7 such diagrams, and there are essentially 3 different Vassiliev invariants of order 4.)

§5 Final Remarks

As we have seen in this chapter, the Alexander-Conway polynomial, the Jones polynomial, and the skein polynomial induce Vassiliev invariants. So, in a sense we may say that Vassiliev invariants are "stronger" than the polynomial invariants. This slightly ambiguous last statement may allow some optimism that the Vassiliev invariants may distinguish two knots. However, at the time writing, this conjecture by Vassiliev remains open. We should emphasize that it is not sufficient just to consider specific Vassiliev invariants. For this conjecture to hold "all" Vassiliev invariants must be taken into account, because we have examples that show there are infinitely many distinct knots that cannot be distinguished by *finitely* many Vassiliev invariants.

In this regard, let K(n; l), $n \geq 2$, $l \geq 0$, be the knot depicted in Figure 15.5.1, where 2l is the number of positive crossing points on the far right band.

K(5;2)

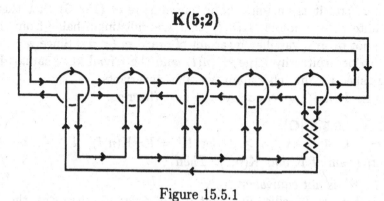

Figure 15.5.1

K(n; 0), $n \geq 2$, is an alternating knot. Since it is possible to find for it an alternating diagram with 3n crossing points, we shall leave this as an exercise for the reader. Therefore. K(n; 0) is not a trivial knot.

The Alexander-Conway polynomial and the Jones polynomial of $K(n; l)$ can be found in the following proposition.

Proposition 15.5.1.

For any integer $l \geq 0$,

$$(1) \ \nabla_{K(n;0)}(z) = \nabla_{K(n;l)}(z)$$

$$(2) \ \nabla_{K(n;0)}(z) = 1 - 2z^n + \ldots \qquad \text{if } n \text{ is even}$$

$$\nabla_{K(n;0)}(z) = 1 - 2z^{n+1} + \ldots \qquad \text{if } n \text{ is odd} \qquad (15.5.1)$$

$$(3) \ V_{K(n;l)}(t) = t^{2l}(V_{K(n;0)}(t) - 1) + 1.$$

Exercise 15.5.1. (1) Prove (15.5.1)(1) and (3).

(2) Confirm that (15.5.1)(2) holds for $K(5; 2)$.

(3) Show that $K(n; l) \approx K(n; l')$ if and only if $l = l'$.

It follows from (15.5.1)(2) for the Vassiliev invariant ∇_m of order m, $2 \leq m < n$,

$$\nabla_m(K(n; l)) = 0. \qquad (15.5.2)$$

The result of (15.5.2) may be extended to *any* Vassiliev invariant, v_m. In fact, it has been proven [O] that for an arbitrary Vassiliev invariant, v_m, of order m, with $2 \leq m < n$,

$$v_m(K(n; l)) = 0. \qquad (15.5.3)$$

Therefore, it is an immediate consequence of (15.5.3) that there are infinitely many knots, $K(n; l)$ that are indistinguishable from the trivial knot by any Vassiliev invariant of order $m \ (< n)$. Since we may take n to be arbitrarily large, $K(n; l)$ and the trivial knot cannot be distinguished by "finitely" many Vassiliev invariants.

Using $K(n; l)$, we can prove the next theorem:

Theorem 15.5.2 [O].

For a knot K and $n \geq 2$, let $K' = K \# K(n; l)$, $l \geq 0$, be the connected sum of K and $K(n; l)$. Then

(1) K' is not equivalent to K;

(2) For any Vassiliev invariant v_m of order m, if $m < n$ then

$$v_m(K') = v_m(K).$$

Therefore, there are infinitely many distinct knots that cannot be distinguished by finitely many Vassiliev invariants.

In the above example K' is not a prime knot. However, there does exist an example of the same property as above in which both K and K' are prime knots [St].

If, on the other hand, we consider "all" Vassiliev invariants, then the situation is quite different. One of the strengths of the theory that surrounds the Vassiliev invariants is that it allows us to treat the polynomial invariants in a systematic way. Hence, the Vassiliev invariants may reveal relationships between the polynomial invariants. In one particular case, a surprising result of this kind has been found [MelM].

Theorem 15.5.3 [BN-G].

Let K and K' be two knots. If for all $N \geq 2$ the Nth degree Jones polynomials are equal, then their Alexander polynomials are also equal.

The above theorem may be paraphrased as follows: The set of all N^{th} degree Jones polynomials $\{\widehat{J}_K^{(N)}(t), N \geq 2\}$ determines the Alexander polynomial. This result in itself is quite surprising, since from Chapter 11 we know that the Jones polynomial, $V_K(t)$, does not determine the Alexander polynomial, $\Delta_K(t)$, and the converse is also true.

The importance of Theorem 15.5.3 is in the fact that it may be a crucial part in the quest to find a non-trivial knot with the trivial Jones polynomial (cf. Chapter 11, Section 2). The equation "$V_K(t) = 1$" may not imply "$\Delta_K(t) = 1$," but as a consequence[18] of Theorem 15.5.3, we now know that

$$\widehat{J}_K^{(N)}(t) = 1, \text{ for all } N \geq 2, \Longrightarrow \Delta_K(t) = 1. \qquad (15.5.4)$$

This may be used as a basis for the following conjecture.

Conjecture. If for all $N \geq 2$, $\widehat{J}_K^{(N)}(t) = 1$, then K is the trivial knot.

So, even if Vassiliev invariants are not strong enough to distinguish two knots, they may be able to distinguish knots from the trivial knot.

As we mentioned in the introduction to this chapter, the study of Vassiliev invariants is still in its nascent stages. But on the basis of present research, we may predict that the Vassiliev invariants will play a major rôle in the future development of knot theory.

For an excellent more detailed introduction to the theory of Vassiliev invariants, we refer the reader to [Bi]. The ambitious reader may care to peruse the more advanced references that may be found in [BN].

Appendix

A table of knots and their knot invariants

Appendix (I) is a complete list of prime knots with up to 8 crossing points. In this set of knots,
(1) The amphicheiral knots are 4_1, 6_3, 8_3, 8_9, 8_{12}, 8_{17}, 8_{18};
(2) There is only one non-invertible knot, 8_{17};
(3) The only non-alternating knots are 8_{19}, 8_{20}, 8_{21};
(4) All the knots are 2-bridge knots, except the following nine knots 8_5, 8_{10}, and $8_{15} - 8_{21}$, which are 3-bridge knots.

Appendix (II) is a table of the Alexander polynomials and the Jones polynomials of the knots in Appendix (I).
The notation
$$(n)[a_0 a_1 a_2 \ldots a_m]$$
denotes the polynomial
$$t^n(a_0 + a_1 t + a_2 t^2 + \ldots + a_m t^m).$$

For example,
$$(-2)[-1 + 2 - 3 + 0 + 4 + 1] = t^{-2}(-1 + 2t - 3t^2 + 0t^3 + 4t^4 + t^5)$$
$$= -t^{-2} + 2t^{-1} - 3 + 4t^2 + t^3.$$

If the Jones polynomial of a knot is given by
$$(n)[a_0 a_1 \ldots a_m],$$
then the Jones polynomial of the mirror image, K^*, of K is given by
$$(-n - m)[a_m a_{m-1} \ldots a_1 a_0].$$

In the case of the Alexander polynomials, the polynomials for K and K^* are the same.

Appendix (I): A table of knots

Appendix (II): Alexander and Jones polynomials

Knot	Alexander polynomial	Jones polynomial
3_1	$(-1)[1-1+1]$	$(1)[1+0+1-1]$
4_1	$(-1)[-1+3-1]$	$(-2)[1-1+1-1+1]$
5_1	$(-2)[1-1+1-1+1]$	$(2)[1+0+1-1+1-1]$
5_2	$(-1)[2-3+2]$	$(1)[1-1+2-1+1-1]$
6_1	$(-1)[-2+5-2]$	$(-2)[1-1+2-2+1-1+1]$
6_2	$(-2)[-1+3-3+3-1]$	$(-1)[1-1+2-2+2-2+1]$
6_3	$(-2)[1-3+5-3+1]$	$(-3)[-1+2-2+3-2+2-1]$
7_1	$(-3)[1-1+1-1+1-1+1]$	$(3)[1+0+1-1+1-1+1-1]$
7_2	$(-1)[3-5+3]$	$(1)[1-1+2-2+2-1+1-1]$
7_3	$(-2)[2-3+3-3+2]$	$(2)[1-1+2-2+3-2+1-1]$
7_4	$(-1)[4-7+4]$	$(1)[1-2+3-2+3-2+1-1]$
7_5	$(-2)[2-4+5-4+2]$	$(2)[1-1+3-3+3-3+2-1]$
7_6	$(-2)[-1+5-7+5-1]$	$(-1)[1-2+3-3+4-3+2-1]$
7_7	$(-2)[1-5+9-5+1]$	$(-3)[-1+3-3+4-4+3-2+1]$
8_1	$(-1)[-3+7-3]$	$(-2)[1-1+2-2+2-2+1-1+1]$
8_2	$(-3)[-1+3-3+3-3+3-1]$	$(0)[1-1+2-2+3-3+2-2+1]$
8_3	$(-1)[-4+9-4]$	$(-4)[1-1+2-3+3-3+2-1+1]$
8_4	$(-2)[-2+5-5+5-2]$	$(-3)[1-1+2-3+3-3+3-2+1]$
8_5	$(-3)[-1+3-4+5-4+3-1]$	$(0)[1-1+3-3+3-4+3-2+1]$
8_6	$(-2)[-2+6-7+6-2]$	$(-1)[1-1+3-4+4-4+3-2+1]$
8_7	$(-3)[1-3+5-5+5-3+1]$	$(-2)[-1+2-2+4-4+4-3+2-1]$
8_8	$(-2)[2-6+9-6+2]$	$(-3)[-1+2-3+5-4+4-3+2-1]$
8_9	$(-3)[-1+3-5+7-5+3-1]$	$(-4)[1-2+3-4+5-4+3-2+1]$
8_{10}	$(-3)[1-3+6-7+6-3+1]$	$(-2)[-1+2-3+5-4+5-4+2-1]$
8_{11}	$(-2)[-2+7-9+7-2]$	$(-1)[1-2+4-4+5-5+3-2+1]$
8_{12}	$(-2)[1-7+13-7+1]$	$(-4)[1-2+4-5+5-5+4-2+1]$
8_{13}	$(-2)[2-7+11-7+2]$	$(-3)[-1+3-4+5-5+5-3+2-1]$
8_{14}	$(-2)[-2+8-11+8-2]$	$(-1)[1-2+4-5+6-5+4-3+1]$
8_{15}	$(-2)[3-8+11-8+3]$	$(2)[1-2+5-5+6-6+4-3+1]$
8_{16}	$(-3)[1-4+8-9+8-4+1]$	$(-2)[-1+3-4+6-6+6-5+3-1]$
8_{17}	$(-3)[-1+4-8+11-8+4-1]$	$(-4)[1-3+5-6+7-6+5-3+1]$
8_{18}	$(-3)[-1+5-10+13-10+5-1]$	$(-4)[1-4+6-7+9-7+6-4+1]$
8_{19}	$(-3)[1-1+0+1+0-1+1]$	$(3)[1+0+1+0+0+1]$
8_{20}	$(-2)[1-2+3-2+1]$	$(-1)[-1+2-1+2-1+1-1]$
8_{21}	$(-2)[-1+4-5+4-1]$	$(1)[2-2+3-3+2-2+1]$

Notes

(1) Combinatorial topology (or PL-topology) is a branch of topology that concerns itself with the study of complexes in Euclidean space. Such complexes consist of points, segments, triangles, tetrahedra, *et cetera*, Figure A.1. These constituent parts are called *simplexes*.

The union of all simplexes contained in a complex C is called a polyhedron and is denoted by |C|. The dimension of a complex is the highest dimension of the simplexes in C. In this book, the simplexes that we shall encounter will not have dimension greater than 3.

A complex C is locally finite, i.e., for each point P there exists a neighbourhood of P that intersects only finitely many simplexes in C.

A knot is a 1-complex (or equivalently, a polygon constructed from 1-complexes) and so by the above definition of a complex a point P of the type in Figure 1.0.1(b) cannot exist on a knot.

For a more detailed discussion of combinatorial topology we refer the reader to, for example Glaser [G*] or Massey [M*].

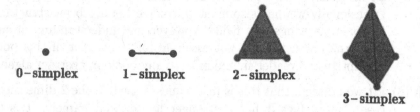

0−simplex   1−simplex   2−simplex   3−simplex

Figure A.1

(2) The set of all knots (or links) that are equivalent to a knot (or link) K is called the *type* of K. Therefore, when we say that the knots K and K' are *equivalent*, we mean that the type of K and the type of K' are *equal*. Therefore, to be precise, knot theory concerns itself with the *type of knots (or links)* rather than the *knot* itself.

(3)   For the purposes of this book we shall deem all maps to be PL-maps, namely, these are continuous maps from a polyhedron |X| to a polyhedron |Y|, which are also linear maps with regard to some division of X. If such a map f is also a homeomorphic map, then f is said to be a PL-homeomorphism. A n-ball B^n is an example of a topological space that is PL-homeomorphic to an n-simplex. In particular, a 2-ball is called a *disk*. While an $(n-1)$-sphere S^{n-1} is PL-homeomorphic to the boundary of an n-ball.

(4)   This is called a one-point compactification of R^2. In general, a one-point compactification of R^n is homeomorphic to S^n, and so it will be more convenient for us to think of S^n as a compact space obtained from R^n by adding a point to R^n.

(5)   Let G be a (non-empty) set. G is called a *group* if for two arbitrary elements a, b in G, we may uniquely define a product ab such that

   (i)   (ab)c = a(bc)   (associative law).

   (ii) There exists a (unique) element e in G such that for each a in G, ea = a and ae = a. We shall call e the identity element.

   (iii) For each element a in G, there exists a (unique) element a* such that aa* = e and a*a = e. We shall call a* the inverse element of a and denote it by a^{-1}.

   In addition, if ab = ba for any two elements a, b in G, then the product is called *commutative* and G is called a commutative group.

   If a set G satisfies (i) [but not necessarily (ii) or (iii)] then G is called a semi-group. For example, the set of all non-zero rational numbers is a group under the usual product, while the set of all integers is not a group, but a semi-group, under the product. Occasionally we may also mention some other algebraic structures, for example, rings and fields; however, an understanding of such (algebraic) structures is not essential for the reader of this book. The interested reader should refer to any book on abstract algebra.

(6)   It would seem that this is just a guess based on the 2-dimensional case, but in fact it has a stronger foundation, namely, it is the Schönflies Theorem.

Schönflies Theorem

   A 2-dimensional PL-sphere S^2 in a 3-sphere S^3 divides S^3 into 2 parts called the interior and the exterior, and each part is PL-homeomorphic to a 3-ball.

   This is the generalization of the famous Jordan Curve Theorem:

Jordan Curve Theorem

A simple closed polygon C in R^2 divides R^2 into two domains G_1 and G_2, namely, $G_1 \cup G_2 = R^2$ and $G_1 \cap G_2 = C$.

We should note that there are examples of 2-spheres that are not PL-2-spheres for which the Schönflies Theorem does not hold.

(7) L_1 consists of two trivial knots but one component of L_2 is a (right-hand) trefoil knot, which we know is not equivalent to a trivial knot. Therefore, L_1 and L_2 cannot be equivalent. To show that their complements are homeomorphic requires a straightforward application of a Dehn surgery, a concept that we shall discuss in Chapter 8.

(8) The root of the difficulty of this problem is that almost all algebraic invariants of knots are powerless when confronted with this problem. H. Trotter had to apply techniques from 2-dimensional hyperbolic geometry to show that the knot in Figure 3.2.2(a) is not invertible. It is not particularly surprising that 2-dimensional hyperbolic geometry is useful in the above case, since 3-dimensional hyperbolic geometry plays a fundamental rôle in the study of 3-dimensional topology.

(9) For a link there may exist a closed polygon β in the interior of $\triangle$. Therefore, an addition to the definition of upper type and lower type may be needed for these β, but this is fairly straightforward from the proof of Lemma 4.1.3.

(10) Let n be a positive integer. We write $a \equiv b \pmod{n}$ or $a - b \equiv 0 \pmod{n}$ if and only if $a - b$ is divisible by n.

(11) **Classification of closed orientable (connected) surfaces**

Let F_m be a 2-sphere with m (≥ 0) handles attached, see Figure 5.2.1(a) or (b). Then

(i) *A closed orientable connected surface is homeomorphic to F_m for some $m \geq 0$;*

(ii) *F_m and F_n are homeomorphic if and only if $m = n$.*

For the classification of non-orientable closed surfaces refer to Massey [M*].

(12) To be exact, these m $(= 2g(F) + \mu(K) - 1)$ closed curves $\alpha_1, \alpha_2 \ldots, \alpha_m$ are chosen so that their homology classes $[\alpha_1]$, $[\alpha_2], \ldots, [\alpha_m]$ form a basis for the 1-dimensional homology group (with coefficients in $\mathbf{Z}$) $H_1(F; \mathbf{Z})$ of a Seifert surface F.

(13)   For two simple closed curves A, B in a Lens space $L(q, r)$, $(q \neq 0)$, one can define the linking number. In this case the linking number $\mathrm{lk}(A, B)$ is not necessarily an integer, it could be a rational number $\frac{s}{q}$ $(0 \leq \frac{s}{q} < 1)$. However, for two simple closed curves in a general 3-manifold, it may not be possible to define a linking number. For example, the linking number between two simple closed curves in $S^2 \times S^1$ cannot be defined.

(14)   The Jones polynomial of an alternating knot (or link) K is also alternating, but Theorem 11.5.2(1) may not hold, and so (2) should be written as $a_i a_{i+1} \leq 0$.

(15)   Since a flat disk is irrelevant to the orientation, a system of flat disks can be defined for any spatial graph, and *flat* equivalence between two spatial graphs may be defined along similar lines to Definition 15.1.1. In fact, for a 3-regular spatial graph there is no distinction between (ordinary) equivalence and flat equivalence.

(16)   Instead of the set of rational numbers, we may use the set of integers or the set of real numbers.

(17)   The set of all Vassiliev invariants of order (at most) m, V_m $(m \geq 2)$, forms a vector space (over the rational numbers). The dimension of V_m / V_{m-1} [the vector space of all Vassiliev invariants of order (exactly) m] is considered as the number of essentially different Vassiliev invariants of order m.

(18)   Equation (15.5.4) is not an immediate consequence of Theorem 15.5.3. A precise relationship between $\{\widehat{J}_K^{(N)}, N \geq 2\}$ and $\Delta_K(t)$ has been established. Theorem 15.5.3 and (15.5.4) are consequences of this relationship.

Bibliography

Books on Knots and Related Topics

[B*] J.S. Birman, **Braids, Links and Mapping Class Groups,** Ann. of Math. Studies 82, Princeton Univ. Press (1974).

[BZ*] G. Burde and H. Zieschang, **Knots,** Studies in Math. 5, Walter de Gruyter (1985).

[CF*] R.H. Crowell and R.H. Fox, **Introduction to Knot Theory,** Graduate Texts in Math. 57, Springer-Verlag (1977).

[G*] L.C. Glaser, **Geometrical Combinatorial Topology vol.1,** Van Nostrand Reinhold (1970).

[HZ*] F. Hirzebruch and D. Zagier, **The Atiyah-Singer Theorem and Elementary Number Theory,** Math. Lecture Series 3, Publish or Perish (1974).

[K*] A. Kawauchi, editor, **A Survey of Knot Theory,** Birkhauser (1996).

[M*] W.S. Massey, **Algebraic Topology: an Introduction,** Graduate Texts in Math. 56, Springer-Verlag (1977).

[R*] D. Rolfsen, **Knots and Links,** Math. Lecture Series 7, Publish or Perish (1976).

Papers on Knots and Related Topics

[Bi] J.S. Birman, *New points of view in knot theory,* Bull. Amer. Math. Soc. **28** (1993) 253–287.

[Br] E.J. Brody, *The topological classification of the lens spaces,* Ann. of Math. **71** (1960) 163–184.

[BL] J.S. Birman and X-S. Lin, *Knot polynomials and Vassiliev's invariants,* Invent. Math. **111** (1993) 225–270.

[BN] D. Bar-Natan, *On the Vassiliev knot invariants,* Topology **34** (1995) 423–472.

[BN-G] D. Bar-Natan and S Garoufalidis, *On the Melvin-Morton-Rozansky conjecture,* preprint, Harvard Univ. (1994).

[BS] C. Bankwitz and H.G. Schumann, *Über Viergeflechte*, Abh. Math. Sem. Univ. Hamburg **10** (1934) 263–284.

[C] J.H. Conway, *An enumeration of knots and links, and some of their algebraic properties*, **Computational Problems in Abstract Algebra**, Pergamon Press (1970) 329–358.

[CG] J.H. Conway and C.McA. Gordon, *Knots and links in spatial graphs*, J. Graph Theory **7** (1983) 445–453.

[D] V.G. Drinfel'd, *Quantum Groups*, **Proceedings of the ICM, Berkeley** (1986) 798–820.

[DT] C.H. Dowker and M.B. Thistlethwaite, *Classification of knot projections*, Topology and its Applications **16** (1983) 19–31.

[ES1] C. Ernst and D.W. Sumners, *The growth of the number of prime knots*, Math. Proc. Cambridge Phil. Soc. **102** (1987) 303–315.

[ES2] C. Ernst and D.W. Sumners, *A calculus for rational tangles; applications to DNA recombination*, Math. Proc. Cambridge Phil. Soc **108** (1990) 489–515.

[F1] R.H. Fox, *A quick trip through knot theory*, **Topology of 3-manifolds and Related Topics**, Prentice-Hall (1962) 120–167.

[F2] R.H. Fox, *Metacyclic invariants of knots and links*, Canad. J. Math. **22** (1970) 193–201.

[FM] R.H. Fox and J.W. Milnor, *Singularities of 2-spheres in 4-space and cobordism of knots*, Osaka J. Math. **3** (1966) 257–267.

[G] C.A. Giller, *A family of links and the Conway calculus*, Trans. Amer. Math. Soc. **270** (1982) 75–109.

[GL] C.McA. Gordon and J. Luecke, *Knots are determined by their complements*, J. Amer. Math. Soc. **2** (1989) 371–415.

[GLM] C.McA. Gordon, R.A. Litherland and K. Murasugi, *Signatures of covering links*, Canad. J. Math. **33** (1981) 381–394.

[H] F. Hosokawa, *On ∇-polynomials of links*, Osaka Math. J. **10** (1958) 273–282.

[J1] V.F.R. Jones, *Hecke algebra representations of braid groups and link polynomials*, Ann. of Math. **126** (1987) 335–388.

[J2] V.F.R. Jones, *On knot invariants related to some statistical mechanical models*, Pacific J. Math. **137** (1989) 311–334.

[Kan] T. Kanenobu, *Examples on polynomial invariants of knots and links*, Math. Ann. **275** (1986) 555-572.

[Kau1] L.H. Kauffman, *State models and the Jones polynomial*, Topology **26** (1987) 395–407.

[Kau2] L.H. Kauffman, *An invariant of regular isotopy*, Trans. Amer. Math. Soc. **318** (1990) 417–471.

[Ki] R. Kirby, *A calculus for framed links in* S^3, Invent. Math. **45** (1978) 35–56.

[KM] P. Kronheimer and T. Mrowka, *Gauge theory for embedded surfaces I*, Topology **32** (1993) 773–826, *II* (*ibid*) **34** (1995) 37–97.

[L] W.B.R. Lickorish, *Three-manifolds and the Temperley-Lieb algebra*, Math. Ann. **290** (1991) 657–670.

[LM1] W.B.R. Lickorish and K.C. Millett, *Some evaluations of link polynomials*, Comment. Math. Helv. **61** (1986) 349–359.

[LM2] W.B.R. Lickorish and K.C. Millett, *The reversing formula for the Jones polynomial*, Pacific J. Math. **124** (1986) 173–176.

[LM3] W.B.R. Lickorish and K.C. Millett, *A polynomial invariant of oriented links*, Topology **26** (1987) 107–141.

[MelM] P.M. Melvin and H.R. Morton, *The coloured Jones function*, Comm. Math. Physics **169** (1995) 501–520.

[Men] W. Menasco, *Closed incompressible surfaces in alternating knot and link complements*, Topology **23** (1984) 37–44.

[MenT] W. Menasco and M. Thistlethwaite, *The classification of alternating links*, Ann. of Math. **138** (1993) 113–171.

[Mi] J.W. Milnor, *Infinite cyclic covers*, **Conference on the Topology of Manifolds,** Prindle, Weber and Schmidt (1968) 115–133.

[Mo] H.R. Morton, *Seifert circles and knot polynomials*, Math. Proc. Cambridge Phil. Soc. **99** (1986) 107–109.

[Muk] H. Murakami, *A recursive calculation of the Arf invariant of a link*, J. Math. Soc. Japan **38** (1986) 335–338.

[Mus1] K. Murasugi, *On the genus of the alternating knots I*, J. Math. Soc. Japan **10** (1958) 94–105, *II* (*ibid*) **10** (1958) 235–248.

[Mus2] K. Murasugi, *On the Alexander polynomial of the alternating knot*, Osaka Math. J. **10** (1958) 181–189.

[Mus3] K. Murasugi, *On a certain numerical invariant of link type*, Trans. Amer. Math. Soc. **117** (1965) 387–422.

[Mus4] K. Murasugi, *On the signature of links*, Topology **9** (1970) 283–298.

[Mus5] K. Murasugi, *On invariants of graphs with application to knot theory*, Trans. Amer. Math. Soc. **314** (1989) 1-49.

[Mus6] K. Murasugi, *On the braid index of alternating links*, Trans. Amer. Math. Soc. **326** (1991) 237-260.

[O] Y. Ohyama, *Vassiliev invariants and similarity of knots*, Proc. Amer. Math. Soc. **123** (1995) 287-291.

[Sc1] H. Schubert, *Die eindeutige Zerlegbarkeit eines Knotens in Primknoten*, Sitzungsber Heidelberger Akad. Wiss. Math. Nat. Kl **3** (1949) 57-104.

[Sc2] H. Schubert, *Über eine numerische Knoteninvariante*, Math. Zeit. **61** (1954) 245-288.

[Sc3] H. Schubert, *Knoten mit zwei Brücken*, Math. Zeit. **65** (1956) 133-170.

[St] T. Stanford, *Braid commutators and Vassiliev invariants*, to appear in Pacific J. Math.

[Tr] H.F. Trotter, *Homology of group systems with applications to knot theory*, Ann. of Math. **76** (1962) 464-498.

[Tu] V.G. Turaev, *The Yang-Baxter equation and invariants of links*, Invent. Math. **92** (1988) 527-553.

[V] V.A. Vassiliev, *Cohomolgy of knot spaces*, **Theory of Singularities and its Applications,** Amer. Math. Soc. (1990) 23-69.

[Wa] J.C. Wang, *DNA topoisomerases*, Scientific American **247** (1982) 94-109.

[Wi] E. Witten, *Quantum field theory and Jones polynomial*, Comm. Math. Physics **121** (1989) 351-399.

[WAD] M. Wadati, Y. Akutsu and T. Deguchi, *Knot theory based on solvable models at criticality*, **Integrable systems in quantum field theory and statistical mechanics,** Academic Press (1989) 193-285.

[Y] S. Yamada, *An invariant of spatial graphs*, J. Graph Theory **13** (1989) 537-551.

Index